Routledge Revivals

The Element of Fire

First published in 1988, the aim of this book can be stated in Nietzsche's words: 'To look at science from the perspective of the artist, but at art from that of life'. The title contests the notions that science alone can provide us with the most objective truth about the world, and that artistic endeavour can produce nothing more valuable than entertainment. O'Hear argues that art and the study of art are not indispensable aspects of human life, and that this is equally as important as the investigation of the natural world.

The Element of Fire

Science, Art and the Human World

Anthony O'Hear

Routledge
Taylor & Francis Group

First published in 1988
by Routledge

This edition first published in 2014 by Routledge
2 Park Square, Milton Park, Abingdon, Oxon, OX14 4RN

Simultaneously published in the USA and Canada
by Routledge
711 Third Avenue, New York, NY 10017

Routledge is an imprint of the Taylor & Francis Group, an informa business

Publisher's Note
The publisher has gone to great lengths to ensure the quality of this reprint but points out that some imperfections in the original copies may be apparent.

Disclaimer
The publisher has made every effort to trace copyright holders and welcomes correspondence from those they have been unable to contact.

A Library of Congress record exists under LC control number: 88000306

ISBN 13: 978-0-415-72775-4 (hbk)
ISBN 13: 978-1-315-85199-0 (ebk)
ISBN 13: 978-0-415-72780-8 (pbk)

The Element of Fire

Science, Art and the Human World

Anthony O'Hear

Routledge
London and New York

First published in 1988 by
Routledge
a division of Routledge, Chapman and Hall
11 New Fetter Lane, London EC4P 4EE

Published in the USA by
Routledge
a division of Routledge, Chapman and Hall, Inc.
29 West 35th Street, New York, NY 10001

Set in 10/12 point Times
by Witwell Ltd, Southport
and printed in Great Britain
by T. J. Press (Padstow) Ltd
Padstow, Cornwall

Library of Congress Cataloging in Publication Data
O'Hear Anthony.
The element of fire: science, art and the human world/Anthony O'Hear.
p. cm.
Bibliography: p.
Includes index.
1. Science–Philosophy. 2. Art–Philosophy. 3. Imagination.
I. Title.
B67.04 1988
001.1–dc19 88–306

British Library Cataloguing in Publication Data
O'Hear, Anthony.
The element of fire: science, art and the human world
1. Civilization. Role of imagination & science
I. Title
909
ISBN 0–415–00618–X

For Natasha, Jacob, Theodora
and Tricia

Artists make mankind an individual by connecting the past with the future in the present. Artists are the higher organ of the soul where the vital spirits of all external humanity join together, and where inner humanity has its primary sphere of action.

Friedrich Schlegel
(*Ideas*, no. 64)

And new philosophy cals all in doubt
The Element of Fire is quite put out;
The Sun is lost, and th'earth, and no man's wit,
Can well direct him, where to look for it.
And freely men confesse, that this world's spent,
When in the Planets and the Firmament
They seek so many new; they see that this
Is crumbled out againe to his Atomis.
'Tis all in pieces, all cohaerence gone:
All just supply, and all Relation:
Prince, Subject, Father, Sonne, are things forgot,
For every man alone thinks he has got
To be a Phoenix, and that there can bee
None of that kind, of which he is, but hee.

John Donne

The First Anniversarie
An Anatomy of the World,
lines 205–217 (1611)

Contents

Preface

The Element of Fire grew out of my inaugural lecture as Professor of Philosophy at Bradford. The lecture was entitled 'On Truth in Science and Art' and some paragraphs from it appear in Chapters 2, 5 and 7 of this book. Various parts of this book have been used in my teaching at Bradford; others have been read at philosophy seminars and societies at the Universities of St Andrews, Dundee, Glasgow and Oxford; at a conference on human nature at the Middlesex Polytechnic in April 1987; at the 1987 conference of the Philosophy and History Section of the British Psychological Society; and at the Open University Arts Faculty seminar. I have benefitted from comments and criticisms made at all these meetings and classes.

I am also very pleased to be able to express my particular gratitude to R. A. D. Grant, Mary Midgley, Betty Redfern and Roger Fellows, who kindly commented on all or part of the manuscript, as did Peter Fuller. I have had many discussions with Peter Fuller and Roger Scruton on the themes of the book, while my thoughts about the Olympia pediment have evolved over the years in conversations with George Pavlopoulos. Mary Hesse was kind enough to let me see and use some unpublished writings of hers, as were John Haldane and Peter Munz, and I am grateful to them as well.

I would finally like to thank Beverley Toulson for her skill and efficiency in typing and preparing the manuscript.

Acknowledgments

For permission to quote from Rilke's First Duino Elegy, lines 26–30, his Tenth Duino Elegy, lines 12–15 and his Sonnet to Orpheus, Book 2, No. 10, lines 1–2, 5–6, from Raine Maria Rilke, *Selected Works, Vol. II*, translated by J. B. Leishman, Copyright © 1960, The Hogarth Press, I am grateful to the estate of the translator, The Hogarth Press and New Directions Publishing Corporation.

For permission to quote R. Lattimore's translation of *The Iliad*, Book 11, lines 172–8, I am grateful to the publishers of the Lattimore Translation of *The Iliad* (1951), the University of Chicago Press.

Introduction

The aim of this book can be stated in the words in which Nietzsche reviewed fourteen years later his original aim in *The Birth of Tragedy*: 'to look at science from the perspective of the artist, but at art from that of life'.[1] This seems to me to be a task well worth undertaking. It is an eminently philosophical task, but one which academic specialisation discourages, for it cuts across and through demarcations and disciplines. It requires us to stand back from a particular demarcation or discipline and ask what its point is in the context of a life and a culture as a whole. This task is one for which our culture, fragmented, doubting, hedonistic as it is, makes us extremely unsuited; it forces on us a type of evaluation which is hard, and which may seem impossible in the absence of a shared framework of belief and order. We must remember, though, the Socratic belief in the worthlessness of the unexamined life. And so, however hard it is, it is worthwhile to examine the relationship of science and art to each other, and to life itself. In doing so, I will contest two of the potent myths of our time: that science alone can provide us with the ultimate, objective truth about things, and that artistic endeavour and expression can in the end produce nothing more truthful or valuable than entertainment and diversion.

These two myths are connected. For if it is science which, uniquely and authoritatively, tells us how things really are in their essence, then other forms of imagination and expression will amount to little more than playful indulgences of our feelings. They tell us nothing of how things really are, and are mere froth on the daydreams of our naive and, from the scientific standpoint, naively distorted conceptions of reality. And who then is to say that one bit of froth has more worth than another? Science will, it seems, show us that human sensory experience itself involves us in all sorts of distortions of the true nature of things, including our own nature, and that in art and aesthetic appreciation we have nothing more substantial than the unwitting exploitation of certain pleasurable neural link-ups.

In the course of the book I will argue that science cannot so radically displace our subjective experience, and that the theories of science must remain tied to the ways things initially appear to us, on pain of losing any title to truth or acceptability. It is true that we are able imaginatively to construct visions of the world which appear to be totally objective and which replace all our familiar and traditional ways of thinking and perceiving. We are also able to investigate the physical world in terms of how it actually is, rather than in terms of its effects on us. But there are inherent limits on how far we can pursue such investigations, without their conclusions taking the form of unprovable myth, as untied from the observable world as an all-embracing ideology or religion.

In no sense is this book 'anti-science'. The ideal of a study of nature, guided by how things are in themselves rather than by how they appear to us or how we would like them to be, is a strongly moral one, involving patience, humility and a respect for how things are. Its first mature elaboration in the sixteenth and seventeenth centuries represented a moral advance on earlier ways of thought in its insistence on objectivity and its painstaking examination as to the meaning of this objectivity and its demands. There should be no turning back from a science guided by considerations of objectivity, at least if we are at all motivated by integrity, honesty or truth. Much of what is now meant by integrity and honesty is inextricably entwined with the practice of scientific research. Modern western science has at its heart a moral imperative we disregard at our peril. Science, indeed, in its intersubjective testing of theory and its submission of theory to test, provides us with the best exemplar we have of integrity and truth. The links between science and democracy, and the unease totalitarian regimes have with truly scientific enquiry are not coincidental.

All this can be said, though, while insisting that science or scientists limit any pretensions they may have to be able to deliver an impersonal – and because impersonal uniquely true – and significant view of the world, in which the human agent and his concerns are displaced as mere by-products of more fundamental processes. I do not say that any particular scientist or philosopher of science actually has such pretensions, though I do believe that pushing the notion of scientific objectivity further than our existence as embodied human beings will properly allow can lead rather naturally to such pretensions. It is for this reason, incidentally, that in my discussions of science I concentrate on the model of physics, for it is in physics that one attempts to arrive at the most general and basic characteristics of reality. It is only natural then that physics and physicists should from time to time tend to suggest a degree of unreality or even illusoriness concerning less general and less basic characteristics of reality. It is just this tendency I would hope to contest. To the extent that it is not prevalent in other branches of science, any criticisms I make of it will not apply there.

It will be obvious that the topics which concern me are those which exercised or should have exercised Snow and Leavis twenty-five or so years ago in their so-called 'Two Cultures' debate. In rejecting the idea that there could be a scientific culture, properly speaking, my sympathies are firmly on the side of Leavis. With Leavis, I will insist that questions of culture are pre-eminently questions of value, of life, and that scientific theorizing, precisely because of its attempt at an impersonal and non-human perspective, can have nothing to say on such matters. Against Snow, I deplore the cultural attitudes consequent upon uncritical faith in the problem-solving powers of technology and its associated political rationalism. Some of what I say in this book about modernism in the arts bears on this latter point. With Leavis, too, I see a crucial role for the arts and the critical–collaborative study of the arts in reaching and expressing well-formed and informed attitudes to our world, and in the formation of what in a characteristically Leavisite phrase we may call the human world. And, like Leavis, I would stress the duality of objective inherited order and individual response involved in the evolution of the human world if it is to succeed in human terms.

On the other hand – and here my sympathies might be on the side of Snow – what I have to say cuts through the disciplinary feuding and sectarianism of the original debate. I have no narrow view as to what might constitute the material which one evaluates or on which we might imaginatively discourse in coming to form our cultural and ethical attitudes. Our Great Tradition includes such cultural achievements as Newtonian physics, the theory of evolution and quantum physics, just as much as Jane Austen, George Eliot and D. H. Lawrence (though it certainly includes them as well). An artist or a critic may well have to come to terms with the scientific achievements of his culture, if his work is to command authority over us. A painter can no longer paint the natural world as Constable did, and as Ruskin believed Turner did, in the conviction that nature is the direct manifestation of the hand of God. And it is largely achievements in science and philosophy which have demystified the world in this way.

On the other hand again, there is a crucial difference between a scientific theory and a work of art or of artistic criticism. The one, but not the other, will be an attempt to express some vision or attitude to the world or some part of it from a consciously human perspective. The one succeeds or fails in terms of the type and quality of response it elicits from its audience, while the theories and classifications of others face the tribunal of a nature which may be quite impervious to the perceptions or interests of human beings. In the course of this book I refer to works of art and artistic expression, intending by that to include works of literature, painting, sculpture, music. But some works of history, anthropology, psychology and philosophy could well fall on the 'artistic' side of the distinction I am drawing, to the extent that they are attempts to elaborate imaginative and involved responses to some aspect of the human world. To the extent that they do they should be evaluated, like works of art, in terms of their truthfulness to the complex fabric of that world in the light of informed human feeling and sentiment. And this, I will suggest, is pre-eminently a matter of practical knowledge rather than of theory.

For what counts here is the way human beings react and feel. If there were an enlightening or adequate theory here – as opposed to the banalizing and reductionist mythologies of psychoanalysis, Marxism, sociobiology and the rest – it would be a theory which was validated in terms of its adequacy to the complex and manifold way in which things are felt to be impinging on us and we on each other. It would be evaluated in the light of the full range and complexity of the *Lebenswelt*, the world of human value and meaning. Feeling, meaning and reaction here are primary: something well appreciated by the builders of the Parthenon, who deviated from strict geometrical form in order that the disposition of columns and entablature should satisfy the eye of the appropriately placed observer, and, by satisfying the eye, instil a sense of the union of the divine and civic orders. Knowledge of the sort involved in such a work can be gained only through experience, and its deployment in practice reflects the accumulated and not always explicitly conscious experience of generations.

In ethics and aesthetics our starting point must always be the values and perceptions we inherit and learn from within our particular culture. It is safe to say, against the pretensions of utopian revolutionaries, that we will often be uncertain as to the precise meaning of a particular value or tradition, or its role in the fabric of our life, until we disturb it in some way. The would-be revaluer of values too often shows no appreciation of the way in which in a society or a culture, successful traditions, institutions and values develop spontaneously, guided not by theoretical considerations but by their often ill-understood interplay with the life and reactions of a community, as a way by which the members of that community come to deal with what Vaclav Havel in a moving attack on utopianism has called 'the scandalous chaos of life and its mysterious fertility'.[2]

There is, in fact, a close relationship between the craving for generality envinced in scientific theorizing and the passions for abstract order and rationally discernible and acceptable goals in human affairs, which motivates the true revolutionary in politics and ethics, as in art. But the quest for explicit theory and total order in human affairs is doomed from the start. This is first because it overlooks the extent to which our lives and perceptions are in the first instance governed by habits, reactions, attitudes, motives and dispositions of which we are often barely conscious and which will often need the insight of a sensitive social critic or novelist to reveal. And secondly, there are insuperable difficulties in forecasting people's reactions to things in advance, which make large-scale planning in human affairs impossible. During the course of this book I will have much to say on those failings of artistic modernism which I see as stemming from a false rationalism in the arts, often in the form of a supposedly 'scientific' approach to aesthetic matters. But it is worth pointing out at the outset that the underlying weakness of rationalism in any field of human activity stems from the failure of rationalists to recognize the primacy of practice over theory in human affairs, and the way in which our explicit knowledge and theory is based in pre-theoretical dispositions, reactions and forms of life.

For, as Wittgenstein argues persuasively in his later philosophy, and as I will suggest at greater length in Chapter Four, effective communication between people presupposes common underlying dispositions to act and react, common interests and needs, common passions and aversions, and a common biology and history. Our explicit beliefs and aims can only be adumbrated and discussed with others on the assumption of much in common between us and our interlocutors, in terms of what we will notice, find significant and agree to. Without a basic community of life and interest, the very terms in which we speak will fail to engage the attention and intellect of our audience. Nor is it the case that any individual communicator creates his own terms or basic meanings. The language in which we express ourselves is itself something given to each one of us pre-theoretically. We each have to learn it as a practice, which means that it will initially be instilled in us and accepted by us on its own terms, rather than as something we accept and assess as some explicit theory we might consider and form an opinion about. Learning a language involves taking on oneself the whole range of values and perceptions embodied in its concepts and classifications. And so the most significant part of each individual's ability to express himself through language is conditional on his having immersed himself in the practice and interest of pre-existing culture.

What goes for language goes for other media of expression. Modernism in the arts, conceived as an attempt to begin music, painting and architecture again from scratch on purely rational or functional premises, was always a non-starter, in that in any form of meaningful expression one presupposes a tradition of existing practice as the background to one's own efforts having expressive force. And the current fashion for 'post-modernism', a sort of superficial eclecticism in which any motif or technique from one past style can be used *ad libitum* in combination with any motifs from any other style, is calculated to reduce the expressive potential of one's work to that of a cultural Disneyland. In post-modernism, the use of stylistic elements is that of the advertiser, who simply plunders the art of past for whatever superficial association he thinks will sell his product; the net effect in both cases is to deprive whatever is produced of any seriousness or effective rootedness in a whole way of life. But art can be more than entertainment or advertisement. Modernism and post-modernism are both to be deplored, in their different ways, for depriving art and artists of the expressive potential which a tradition of artistic practice and its roots in the life of a people can make available for us. The attack on modernism carried on in this book is no apologia for post-modernism, but rather an exhortation to artists to rediscover and by rediscovering to revivify the artistic traditions to which we are heirs.

It is a strange paradox that healthy continuity and organic growth of tradition is currently far more clearly evident in the scientific realm than in the artistic, but the absence of shared symbolic and practical orders is as destructive in the field of artistic endeavour as it would be in the scientific. There is a lot, humanly speaking, that artists could learn from the way in which scientists unreflectively initiate their successors into existing traditions of investigation and research. Not to do this in science would seem simply an unserious approach to a serious matter and imply that any beginner could recover for himself the achievements of generations. But, as I hope to show, the development of the human spirit in artistic endeavour is at least as serious a matter for a culture as the investigation of the natural world.

CHAPTER 1

The Scientific and the Artistic: Two Approaches to Experience

In the First Duino Elegy, Rilke wrote:

> Many a star
> was waiting for you to perceive it. Many a wave
> would rise in the past towards you; or else, perhaps,
> as you went by an open window, a violin
> would be utterly giving itself. All this was commission.
> But were you equal to it?

Many philosophers will, I imagine, reject and some even be repelled by what they see as the extraordinary anthropocentrism of the first two lines I have quoted.[1] During the course of this book I will argue first that there is a sense in which an anthropocentric view of the world is inevitable for us, and that this is not something we should deplore or attempt to sidestep in an effort to ascend to what might be regarded as a more objective viewpoint. Then, secondly, I want to explore, and by exploring to defend, Rilke's conception of human beings making themselves equal to what they perceive, particularly when what they perceive is, like a violin 'giving itself', a human work. In this chapter, however, I shall be mainly concerned to show some of the differences between the vision of the world in modern science, in which reference to human perspectives, types of perception and concern is intentionally eliminated, and the world of human culture which inevitably takes the human being as its centre and focus.

The vault of heaven

One stands, let us say, on a clear night beneath a moonless sky in the Mediterranean, and the sense that there is a vault above one filled with stars is overwhelming. One *knows* that there is no such thing, that the sky and heaven's vault are illusions generated by the peculiar and particular circumstances in which one is observing what one sees; one knows, too, that the particular configurations and constellations of stars one sees have no cosmic or astronomical significance; and even, perhaps, that some of what one is seeing no longer exists, that at least some of the light stimulating one's retina now comes from bodies that ceased to exist long ago. These thoughts, together with the knowledge that in other parts even of this world, such as Australia, say, the heavens would look rather different, might well incline one to the conclusion that objectively there is no such thing as heaven's vault and that one's perception of the universe as a vault above one's head is simply due to highly subjective factors. So far as it goes, this would be a correct inference, but we must be careful just what further conclusions we draw. What is overlooked in the thesis that the vault of heaven is just subjective illusion is the fact that anyone observing the heavens visually would be bound to see himself surrounded by stars and other heavenly bodies which appeared to be on the inside of a sphere or hemisphere that was enclosing him. If he were situated on a large body, such as the earth, he would, in other words, see the heaven as a vault. This is due to the convergence of light from all the visible stars on the sensory organs of a particular recipient, who perceives the stars as spreading out from him on the inner surface of a vault that surrounds him. If the vault of heaven is an illusion, then, it is an illusion which is part of any visually conscious perceiver's picture of the heavens. That any consciousness of the appropriate sort should have such a perception should make us pause before we consign the vault of heaven to the place in our philosophical framework reserved for 'mere' illusions.

I do not want to claim, with Rilke, that the stars were waiting to be perceived by us in the way we perceive them. But something which belongs, and necessarily belongs, to the way in which certain sentient beings perceive some physical reality, surely has for that very reason a perfectly good title to be thought of as part of what the world is really like.

The meaning of this claim can be brought out by contrasting it with the opposing view, which moves from apparently unassailable scientific premises about the nature of what we perceive to the conclusion that the world is in its true nature unperceivable.

The premises which enabled me to cast doubt on the vault of heaven were all derived in one way or another from the realisation that our perceiving the heavens as a vault and the particular nature of the vault we perceive at any given place or time is due rather more to the accidental and contingent circumstances of our perception than to what might be thought of as the essential natures and interrelationships of the bodies we perceive. In particular, we perceive bodies as close to each other and at the same distance from us, when they are actually immeasurably distant from each other and not connected at all. We feel that a more objective viewpoint would show this, and would also somehow dissolve the idea of a heavenly vault into the trackless and endless depths which actually constitute the heavens. But while we may feel this, we need to ask whether the heavens could ever be *perceived* in this way. Wouldn't any perceiver, or any perceiver with visual capacities remotely like ours, always perceive the stars as being points on an invisible dome that surrounded him? Perhaps the more objective viewpoint turns out not to be a possible viewpoint at all, but a conception rather more abstract than anything that can be seen.

Scientific objectivity

There are, naturally, good reasons for adopting the more abstract conception, particularly if one is interested in examining the causal relationships that exist between the heavenly bodies. Mercury may appear to be moving through the constellation Sagittarius at the moment of someone's birth, but astronomy teaches us that the planet and the stars in the constellation have no causal relationships of any significance whatever and appear to be related only because of the way we perceive them. The constellation and Mercury's movement through it are, in other words, a pure phenomenon; the mistake of astrology is to treat pure phenomena as if they were causally significant.

Science deals in causes, in what produces physical effects in the world. The laws and classifications of science are drawn up to encapsulate what we know of causes and causality. Two objects or substances that look very different may be classified together for scientific (= causal) purposes, for example limestone and marble; conversely stuff that appears similar, and is exploited by builders, say, for its manifest similarity may, on scientific classification, be very different. An example here, to which we shall return, would be what stone masons refer to as ornamental marbles, which include onyx, porphyry and marble itself, but exclude limestone.[2] The stones that constitute this set are at the level of chemical structure – at the level of deep causality in other words – very different, one an oxide, one a silicate and marble itself a carbonate, and will in certain circumstances behave differently despite their apparent or manifest similarity.

Modern science, by which I mean science since the turn of the 17th century – when Galileo taught that science was to deal with the primary qualities of things, such as shape, size, number, position and 'quantity of motion', rather than their appearances to us – has tended to undermine the standing of the manifest image of the world in ways more significant than those we have already touched on. As is well known from Eddington's famous discussion of his two tables, modern physics teaches that the table on which Eddington the author is writing – solid, static, coloured and smooth – is, in reality, no such things, but 'mostly emptiness' a collection of rapidly moving and invisible electric charges separated by enormous relative distances, insubstantial, and possessing neither colour nor texture.[3] If what Eddington says is true, then *the image* of the world given to us in science is one we could not perceive at all. It is not just that the particles that go to make up the table are too small to be perceived. Any particle that we were able to perceive, we would perceive as being of such and such a colour and having such and such a texture and smell. Yet it is just these properties that the particles of the physicist lack. So we reach the paradoxical position that even if we were able to perceive the particles that make up the table, we would still not be perceiving them, as they really are. We will at most be perceiving their effects, as mediated and distorted by our perceptual faculties, much in the same way as we naturally see the heavens as a vault.

Actually, the situation is far worse than Eddington envisaged. Standard quantum theory tells us that particles exist which have position but no momenta, or momenta but no position, and which can jump from one position to another without going through the intermediate positions; and also that there can be systems of particles that are in a number of states at one time without being definitely in any single one. (This last is the phenomenon of superposition). It is very hard to see how we could form an image of any of these things, let alone perceive a world that consisted of such things. As Bas van Fraassen puts it:

> once atoms had no color; now they also have no shape, place or volume.... There is a reason why metaphysics sounds so passé, so *vieux jeu* today; for intellectually challenging perplexities and paradoxes it has been far surpassed by theoretical science. Do the concepts of the Trinity, the soul, haecceity, universals, prime matter, and potentiality baffle you? They pale beside the unimaginable otherness of closed space-times, event-horizons, EPR correlations and bootstrap models.[4]

Van Fraassen goes on to say that only someone who does not follow the theologian in saying '*Credo ut intelligam*' (I believe *in order that* I may understand) would not be baffled by a desire, such as van Fraassen has, to limit belief to what can be disclosed in experience. (We find an example of just such belief *ut intelligat* in Professor J. C. Polkinghorne's *The Quantum World*, where we are advised to live with the paradoxes of quantum theory, and submit to a period of 'subliminal absorption' in order to get over their strangeness!).[5] We are fast coming to the conclusion that the scientific image of the world is not an image at all; it is not something that a perceiving consciousness such as ours could experience, or even imagine, if by imagine one means construct a visual image of; this thought is reinforced further if we include as part of the scientific image, as many would, the idea that the passing of time itself is simply an aspect of the observer's experience, and that if we looked at the world with total scientific objectivity, we would see it as a four-dimensional block, a relativistic space-time manifold, with temporal relationships as dependent on spatial ones, and we ourselves as space-time worms embedded in it like fossils.

I said earlier that science aims at the discovery of causes. In doing so, it begins at once to move away from the appearances of things towards what would naturally be called a more objective standpoint, more objective because the causal effects of, say, a piece of marble and a piece of limestone, apart from the way they appear to the senses, would be the same for any creature or object in the universe, quite independently of whether it perceived marble and limestone as being of similar appearance or not. But 'objective' also has another connotation, and here we have to be more cautious. This is the sense in which we speak of an objective view as being one which is not flawed by bias or prejudice on the part of the perceiver, and hence that a more objective account of a state of affairs is generally to be *preferred* to a more subjective one. A more objective view may be said to be one that aims to command the assent of more participants or observers, and ultimately to prescind from the notion of observers and participants altogether, in giving an account of how the world is in itself, independently of how any perceivers or groups of perceivers perceive it.

It is hard to see how any embodied consciousness, let alone a human being, could perceive the world except from a specific time and place; hence it could not perceive the world timelessly, with all times and places equally present and equally distant. If it had vision, it could hardly avoid perceiving the heavenly vault, and even if it did not perceive colour it would still see shade. Similarly, if it had hearing it would perceive sound, touch would reveal texture, the nose odours and the tongue taste. But all these qualities which are the immediate and proper objects of our senses are just the qualities which the physicist's account rules out as being secondary, not really in the objects themselves, and this may be no coincidence. If a creature is able to sense things in the world at all, there must be some mechanism which establishes a link between some particulars in its environment and the subjective mental constitution of the perceiver. Sensory qualities are that link, the means by which things in the world affect and stimulate our minds, meshing with something already in us so as to attract our attention in particular ways.[6] It is because they are partly the result of our make up, that physics has been inclined to deny reality to sensory qualities, such as colour, sound, taste, smell and felt texture, explaining our perception of colour, for example, in terms of the interaction of colourless light rays with our visual cones.

Any sensory contact with one's environment presupposes some mechanism in oneself which is capable of being stimulated by things in the environment, and initially to organise the environment in terms of the way it meshes with those particular sensitivities in one's own make-up which we call our sense organs. It is precisely because these sensitivities in our make-up do not mesh exactly with the most important causal properties of things in the world that we are led to think of the image our senses give us of the world as being in some philosophical sense illusory, in not revealing things as they are in themselves; how they are in themselves is what the physicists' account purports to tell us, for in that account our sensations and perceptions are taken to be secondary rather than primary qualities of things, and the world centering on us and our sensory modalities largely an illusion due to those modalities.

For a consciousness, on the other hand, the world centres on it and is perceived in terms of its senses. It may be possible for certain purposes to adopt a more decentered viewpoint, to reject the concept of the vault of heaven and to prescind from one's modes of sensation, but so long as one stays embodied and conscious, this is how the world will appear. I take all this to be a necessary corollary to one's being embodied and conscious, and this must be the start of any resistance to the attempt to downplay or denigrate the appearance which the world presents to us as being really dispensable or illusory or always to be rejected in favour of the more objective version. The stars were not waiting for us to perceive them but only we could perceive them in the way we do. Only in us, with our particular type of sensibility and intelligence, could this particular aspect of the universe be revealed, could light be thrown on the universe in *this* way. This fact alone and its great antecedent improbability if simply the result of chance has seemed to some to be cosmologically significant, leading to the thought that the universe had necessarily to produce conscious and intelligent beings such as ourselves.[7] Whether our existence is cosmologically significant, however – and perhaps *we* are not in a good position to judge on such a thing – it is certainly significant in human terms which is, perhaps, the only type of significance we are properly qualified to pronounce on. In chapters 2 and 3 I will explain why a certain degree of scepticism is in order regarding the pretensions of science to deliver truths about the world as it really is, as opposed to how it appears to us, but for the moment I want only to contrast the conception of the world which modern science aims at with what, for a better phrase, I will call the human world.

Pure realism: the world as it is

What modern science aims to do is, in the name of a wider objectivity of view, to displace the human being and his modes of perceiving the world from the centre of the picture, and to present the human being and his modes of perceiving the world as incidental parts of the picture. From this point of view, the human being is seen as part of a wider and more inclusive causal process, and, from that point of view, of no more significance than any other incident in the development of the cosmos. Wittgenstein gave expression to what might seem to be the emotional and ethical corollaries of this perspective (or, more properly, non-perspective), which he called 'pure realism', when he wrote in his *Notebooks* in 1916:

> The human body, however, my body in particular, is a part of the world among others, among animals, plants, stones, etc. etc.
> Whoever realises this will not want to procure a pre-eminent place for his own body or for the human body.
> He will regard humans and animals quite naively as objects which are similar and which belong together.[8]

What one is being invited to do here is to step outside one's own life, one's feelings and one's experience; in a sense, to see no value in them, to see them from outside, as it were, as if they were merely organic or even inorganic processes, and subject to the same sorts of causal processes as such processes. In the same way, though less radically, some sociologists attempt to demythologise particular values we hold regarding our traditions and institutions, interpreting them as simply the causal effect of various (usually self-interested) social forces rather than as values that we ought to accept and ought to motivate us.

We find another and intriguing example of the same sort of distancing from what we pre-theoretically feel in some of Wittgenstein's own remarks on proof in mathematics, in which he appears to regard the sense of compulsion with which a proof appears to us to follow from its premises as no more than an aspect of the anthropological fact that we *do* regard certain strings of formulae as proofs.[9] The inward sense we have of something *having* to follow from something else is thus seen in terms of our minds simply being caused to reason in this way by genetic and social factors, inviting the thought that with a different set of causal factors determining our psychology and sociology, different calculations might have appeared compelling to us. But, of course, that thought is one a mathematician doing mathematics cannot have, for it destroys the notion that a proof in mathematics points to something that cannot and could not be different; for the mathematician the thought that 2 + 2 might not equal 4 must be regarded, in Frege's terms, as a 'hitherto unknown kind of insanity'.[10] In all these cases, there is a reinterpretation of something we immediately feel or believe from within in terms of causal factors outside our control, and a consequent tendency to displace and devalue the feeling itself and to cease to see it as compelling on each one of us. Why regarding our feelings about what matters to us in causal terms in this way should lead to a devaluing of these feelings may not be altogether clear; but in opening a gap between ourselves and what we feel, it often does.

However, in so far as we are alive, we do feel about things from within; some things matter to us, and are not simply on a par with all other causal processes. The sociological attitude is not something we can adopt consistently to everything about our lives and our values, nor can we consistently view our bodies and those around us in the way Wittgenstein apparently recommended in 1916. To attempt, *per impossibile*, to adopt such a detached view of one's own life and values and to the other people around one may turn one into some sort of saint; for most of us more likely it will serve only to poison that sense of identification with one's relationships and values which is the basis of a life lived with meaning and a sense of fulfilment.

What concerns us here is a particular aspect of the familiar distinction between the world as it appears to one and the world as it is in itself. The world as it appears to me is a world that appears to an individual consciousness, and does so in a particular way, determined both by particular sensory qualities and by the manner in which things stimulate me, engage me, attract me, repel me. There is a sense in which I am necessarily the centre of the world of appearance. The world of appearance is a world centred on the perceiving ego; the vault of heaven surrounds *me*, the stars appear to *me*; things matter or fail to matter to *me*, initially in so far as and in the manner in which they stimulate my senses, and attract or repel me. The world as it is in itself may be unlike the I-centred world of appearance in all sorts of ways. Absolutely speaking, there is no vault of heaven; absolutely speaking, my pleasures and pains do not occupy the centre of anything, because, absolutely speaking, there is no centre; my pleasures and pains have no significance for the cosmos as a whole except in so far as they are part of the causal history of one organism among many others on one planet among many others.

The world of appearance is a world in which there is significance or mattering for the consciousness who is appeared to, or at least the possibility of such significance. The world as it is in itself, by contrast, denotes a conception of the world in which one is detached from such significance. In particular, it denotes a conception of the world which is impersonal, requiring for its elaboration no concepts which are simply features of the mode of perception peculiar to creatures of a certain physiological type. It refers instead only to those features of reality which are thought to form the underlying basis for all the different actual disclosures that are made to perceivers of different types, and which, in their purity and abstractness, are actually perceived by none. If colour vision is such a feature of the mode of perception of a certain terrestrial species, then the notion of colour will play no essential role in an account of the processes governing the world as it is in itself. At most, it will come in when the behaviour of colour-perceiving beings is to be explained: 'When stimulated by light rays of such-and-such a wavelength in such and such conditions, these creatures tend to stop; (and this they call perceiving a red traffic light).'

One might well question how complete an account of the world as it is in itself, which referred to sensory properties only in this oblique way, could be, given that our conscious states and our idiosyncratic perspectives on the world are facts about the world. And this, I take it, is the essence of the dispute between physicalists (materialist philosophers who regard the mind and mental events as in some way identical with the brain and its states) and their opponents, a dispute much muddied on the physicalist side by an over-free use of the term 'representation'. In physicalist hands, 'representation' can stand *both* for my perception-experience of, say, the mist outside, which is what I enjoy, feel, am made apprehensive by, and so on, *and* for whatever brain-event is caused by the mist and causes me, say, to drive a bit slower. But, it is dubious that my consciously seeing the mist is to be analysed as my having a *representation* of the mist, if a representation is an image or message of some sort in my mind or brain. Certainly, in perceiving the mist I do not scrutinize something in my brain, which I then compare with the mist, although much scientific talk of perception has just this type of implication. One immediate problem here is to ask just what or who (in my brain) does the scrutinizing or comparing.

Significantly, and perhaps to avoid just this type of question, and its suggestion that there might be a homunculus or little man in my brain doing the scrutinizing and comparing, Patricia Smith Churchland, a leading exponent of the physicalist point of view, actually explains the thesis that the brain 'represents' as meaning that there are 'systematic relations between events in nervous networks and other events, some of which are external to the brain and some of which are internal, such that the nervous system can behave adaptively'.[11] It is, then, hardly surprising that a little further on she speaks of a voltmeter 'representing' the voltage across a wire, or the response pattern of a cell population in the striate cortex 'representing' a bar of light. But this type of unconscious representation completely glosses over the particular way in which in our experience the world is revealed or disclosed to consciousness, and the value people put on aesthetic and other types of experience becomes quite mysterious.

Whatever one might say about physicalism in general, it is significant that we, as human beings, are capable of envisaging an account of the world which is not centred on the perceiver, and which abstracts from our particular ways of perceiving the world; and, further, that in modern science we have gone some way to producing such an account; notwithstanding doubts about the distance one can actually go in this direction, what has been done here is a testament to the range and power of human imagination. We are, then, capable of imaginatively transcending the sensations of the present moment, in order to frame a wider, more inclusive, and more objective account of the world, in which our lives and perceptions are seen as the result of wider causal processes, and in which some central aspects of our experience are regarded as secondary and subjective.

Imagination and subjectivity

But how far does the imaginative ability we have to transcend our present sensations in some form of discursive or expressive activity inevitably lead us away from our embodiment and from what concerns us specifically as human beings? Are there not modes of expression and discourse in which the aim is not to transcend one's embodiment and the perspective that goes with that, but rather to explore one's embodiment and perspective as it were *from within*, and at the same time to express this sense objectively, for others to examine? Could there be a realm of expressive activity that is at once objective and subjective? Reflection on human culture should make it obvious that there is such a realm, which consists neither of spontaneous, natural expressions of experiences and sensations, nor does it engage in that type of reflection which would seek to abstract from our embodiment altogether in order to provide an observer-independent view of the world, such as we see in science. It is rather that which results from attempts to classify, describe, order and find meaning in and through our existence and embodied consciousness and centering on the experience we have of the world, and how things appear to us. As such it remains an expression and projection of an anthropocentric picture, in contrast to science which attempts to describe the world from no point of view at all.

Artistic activity is certainly an important part of this type of expression, and in some ways the most personal and active one but it is by no means the only one, nor is it the most fundamental one. For that we would have to look to language itself, or at least to that part of language which is concerned with the expression and evocation of inner human feelings and attitudes. It is true that the expression of the sort of human significance I am interested in here cannot be divorced from the institutions and practices of a culture more generally. For a culture is that public realm in which human beings express their feelings, attitudes and concerns objectively, rather than by mere animal noise, and execute and further their purposes institutionally and socially rather than by instinctive activity, and much of what one expresses and many of one's attitudes and desires will derive from one's cultural background. However, in order to further the contrast with scientific *discourse*, what I will be primarily concerned with is artistic expression (including, of course, literature), for here we have a realm of discourse and expressive activity, which, like science, treats *imaginative* description and expression as *an end in itself*, but in contrast to science, focuses explicitly on the imaginative exploitation of human perspectives and concerns. In so far as a work of artistic expression is regarded aesthetically (which is part of what is involved in seeing it as a work of imagination, produced as an end in itself), it will be contemplated rather than consumed or destroyed. But in so far as such a work is itself intended to be an object of *experience*, relating to human perspectives and concerns rather than prescinding from them as science does, like science it will be intended to arouse in the perceiver attitudes such as attraction, repulsion, formal and intellectual interest and even, on occasion, involve moral assessment. Unlike a scientific theory, its success or failure should be assessed in terms of the responses it evokes.

The language of feelings

Artistic expression stands between the natural unreflective expressions of feeling which are to all intents and purposes part of our physiology, and the type of scientific description which attempts to prescind altogether from our experience of the world. Like the language we use to speak about our feelings and sensations, it is a realm in which *articulated* expression and description of experience or attitude are of the essence. Reflection on the language of sensations can help to bring out one important aspect of artistic activity. In language, an expression of feeling is detached from the subjective feeling or reaction which gives rise to it, and is presented as an object for conscious scrutiny. The expression in question, accordingly, detaches its originator from immersion in the immediacy of his feeling or experience, and in becoming an object for him and for others immediately raises a question as to its adequacy as an expression or description of the experience. The key point to grasp here is that this demand for adequacy means that the originator cannot be content with the expression simply *seeming* to him to be adequate. If one is interested in adequacy of expression, there must be room for a distinction between what really is adequate, and what merely seems to one to be adequate, and this in turn will mean the possibility of judging what is expressed by objective standards, standards, that is, which are in principle applicable by others. To refuse to listen to criticism from others when one is attempting to articulate in words or some other medium an experience or attitude one has is tantamount to saying that one has no interest in the meaning of what one is expressing or in its expressive adequacy. One is simply, in Hegelian terms, setting one's will against that of others; but even if one succeeds in crushing the wills of those around one, one does not thereby succeed in objectively articulating one's experience or vision, even for oneself, nor does one achieve the unforced recognition of one's consciousness which one is looking for from those other wills.

What I have just been saying has both Hegelian and Wittgensteinian roots. Both these philosophers insist that when someone wants to say how it is with him, rather than simply emitting a cry of pain or a sigh of pleasure, he has to use a public language, whose words are intelligible to others, to make judgements that are criticizable in terms of standards available to others. Only against such a background will one's words and judgements become more than inarticulate gibbering. One's words must be used consistently, to express judgements that are objectively correct. Consistency and objectivity imply interpersonal standards in use and judgement, and standards which cannot remain the property of one consciousness alone. The arguments of Hegel and Wittgenstein apply directly to the language in which one refers to one's sensations. As an example of what I mean here, consider the use of the expression 'I have a pain in my foot' as opposed to emitting a cry of pain. In learning this expression, the child has to learn the circumstances of its use, the type of sensation to which it is applied, and some idea of the varying intensities of different pains. He is learning to give objective expression to subjective experience, to locate his experiences in the framework of public, interpersonal discourse; although fully understanding such discourse on the part of others requires from them some inner understanding of what it is like to be in pain. In this movement from inwardness to outwardness and back to inwardness, the language of pain can serve as a model for what I am calling artistic expression, even though artistic expression is disinterested in ways typical expressions of sensation are not: an artist is not through his art seeking some immediate solace or response to his condition as a result of his expression, even where his art could be regarded as an expression of his own inner state of mind, and of course not all art can be so regarded (although I think some can).

Artistic expression and the human world

Much artistic expression, from the Homeric use of metaphor on, is concerned with expressing a possible and revealing way of experiencing the natural world not as science does, but as something presenting a specific appearance to an observer; so it is, in this respect, wider than the language of sensations. Moreover, the realm of artistic expression is wider than the language of sensations in another important respect. Aesthetic expression is not confined to language, but includes other media of expression. These other media and the meanings they convey are not governed by grammars and dictionaries in the same way as language.[12] Nevertheless, there is a surrogate for linguistic consistency in what a tradition of, say, music or architecture has established. What someone in a given tradition expresses is determined not wholly by his individual whim, but in larger part by the conventions and meanings established in his tradition, even and perhaps especially where he goes against the normal expectations or rules. Still, artistic expression can be regarded as an extension of simple linguistic expressions of states of mind in that all artistic expression or thought is rooted in our nature as centres of consciousness embodied in certain specific ways and has the same combination of inner and outer aspects, moving by means of objective expression from what is personal to one to what is personal to another. As Adrian Stokes put it, 'all poetry, all art, all culture, are projections of egocentric attitude',[13] meaning that they all take human feelings and human response as their point of reference, and, I would add, their point of return. This goes for those types of art which have, as their ostensible object, depictions of the natural world, as much as for something like Tchaikovsky's later symphonic music which may seem to the listener to move without the mediation of any depicted object straight from the composer's feelings to his own, from the heart to the heart, as Beethoven put it in his epigraph to the *Missa Solemnis*. Homer's metaphors in Book III of the *Iliad* will work for me only if I can understand what it means for Menelaos to have a sensation like that of a lion chancing in his hunger on a mighty carcass, as of a horned stag or wild goat, when he comes face to face with Paris; and only if I can sense what it is like for Paris to react to Menelaos as a man coming on a snake in a mountain valley, suddenly stepping back, shivers coming over his body. Equally if I do not almost feel the swell of the sea in Turner's *A Ship Aground* (Tate Gallery), the canvas will remain inert for me, like a merely photographic record of a scene. Most works of art either express some human feeling, mood or attitude, or give expression to some actual or possible human experience or evaluation, though not necessarily those of the author; they deal in human feelings, and work (or not) through appeal to the senses and sensibilities of their audience. In this they are unlike a scientific theory, which works if (and only if) its predictions are true, or a piece of technology whose success is again quite independent of audience response, and needs not take human experience as the fundamental point of reference.

What I am calling the artistic realm is a crucial part of what philosophers in the phenomenological tradition have called the *Lebenswelt* – the world of human meaning, institutions and feelings – and which F. R. Leavis called the 'third realm':

> the collaboratively created human world, the realm of what is *neither* public in the sense of belonging to science (it can't be weighed or tripped over or brought into the laboratory or pointed to) *nor* merely private and personal (consider the nature of a language, of the language we can't do without – and literature is a manifestation of language).[14]

It is a public realm, but its publicity depends on a sharing of response, a sharing of values, and most important, a shared understanding of the possibilities of modes of expression, and the values and outlooks embodied in those modes of expression. For our feelings, experiences and attitudes are not simply there, prior to any acculturation. The states of mind we can have are, to a large extent, given to us by the language, the artistic and cultural heritage and social institutions which go to make up the third realm. This is obviously the case with an attitude like patriotism, which depends for its possibility on the existence of a nation, but is also true of emotions like shame or envy, which are the particular forms of discomfort they are only through the subject having the idea of his failing to meet certain standards necessary to his self-esteem or resenting someone else's success. It is hard to see an animal or even a child without some linguistic sophistication being able to capture the precise nuances which make these attitudes differ from, say, guilt or jealousy. Indeed, such an achievement is impossible.

Mention of the way in which human language and institutions actually create the possibility of attitudes and emotions should also serve to correct any impression which may have been given in our speaking earlier of the adequate articulation of experience and in concentrating on the example of the language of pain, as if the experience expressed always exists fully formed and simply needs the right form of description applied to it. Much language and many works of art actually create new forms of experience and attitude. We can see this by considering the creative use of metaphor in poetry or the highlighting of some specific feature of a scene in a painting; the adequacy in question may not be the adequacy of correspondence, as between a map and a piece of terrain, so much as an adequacy and desirability of a creative *response* to some complex situation, or the adequacy with which a new vision throws new light on some subject matter. Ruskin points out that before Turner no painter had depicted a shadow as scarlet, as counterpoint to white sunlight.[15] Had anyone seen shadow as tinged with scarlet before Turner? Looking at, say, *The Decline of the Carthaginian Empire* (Tate Gallery) or *Modern Italy, the Pifferari* (Glasgow City Art Gallery) can anyone fail to experience such a scene in this way? Notice, incidentally, how the essence of these experiences of intense sunlight are missed by supposedly objectively accurate photographs. As one critic puts it, for Turner 'the "meaning" of the picture was in its truth to nature – that is, in its accessibility to everyone as a mimetic "reminder" of the experienced world.'[16] We can understand what Turner's 'truth to nature' means by contrasting Turner's visions of Venice with the almost obsessive unnatural-naturalism of a Canaletto. Despite a profound human absence in Turner's vision, he, unlike Canaletto, is presenting us with a vision of the world which illuminates and expands the experience of the onlooker. It is not, though, only romantic art which works on our experience by means of a subjective 'truth to nature'. The marvellous sense of dancing and movement conveyed by the figures in Agostino di Duccio's low reliefs in the Malatesta Tempio in Rimini is achieved by the unnaturalism of showing the legs in profile and their torsos frontally, though this does not appear to the perceiver as a distortion. To the perceiver it is true to nature, true to the way nature appears.

The nature of this relationship between works of artistic expression, and what they are attempting to express or elaborate, as well as their foundation in and appeal to egocentric – or perhaps better – anthropocentric attitudes explains a feature of critical discourse which was noticed by Leavis, but not I think fully explained by him. With a map or scientific theory we can think of objectivity in terms of correspondence between the artefact and existing states of affairs. Quantification within the artefact – scale and detail in a map, greater precision of prediction within a theory – can enhance possibilities of more precise correspondence. With a work from the cultural realm, on the other hand, an element of subjectivity is always introduced through the reference and appeal to human response. But if it were mere subjectivity, then we would be back with inarticulate cries and other unreflective, instinctive behavioural responses. So what expressive language generally, and artistic works in particular, introduce is the possibility of a shared subjectivity, in which one's inwardly felt reactions are formed through entry into a human cultural world, a world in which the reactions of others to their world have over time been crystallised through objective articulation. Structure and content are thus given to our inchoate and unreflective feelings and attitudes by a process of interiorizing our cultural heritage (as Polanyi would have put it); but in so far as what is at issue is our own response, what we add and necessarily add to what is given to us from outside by language and culture is our own felt agreement or disagreement with some part of what we are given, and our own felt reactions to what we perceive around us.

In so far as we express these reactions objectively, they may or may not awaken a response on the part of others. Our judgment, at any rate, takes the form of an implicit 'This is so, isn't it?', which as Leavis says is an appeal for confirmation of our personal reaction that the thing *is* so

> though expecting, characteristically, an answer of the form, 'yes, but –' the 'but' standing for qualifications, reserves, corrections. Here we have a diagram of the collaborative-creative process in which (a) poem comes to be established as something 'out there', of common access in what is in some sense a public world. It gives us, too, the nature of the existence of English literature, a living whole that can have its life only in the living present, in the creative response of individuals, who collaboratively renew and perpetuate what they participate in – a cultural community or consciousness. More, it gives us the nature in general of what I have called the 'third realm' to which all that makes us human belongs.[17]

The objectivity of a critical judgment of a work of art, then, is an objectivity rooted in the subjective feelings of individuals. What becomes objectively established is what individuals find themselves responding to rather than what is 'proved' in some external, objective fashion. What is responded to in turn structures the world and feelings of succeeding generations, who then build on what they have been given by continuing the critical–collaborative process. It is in this light that new works of art can themselves be regarded as part of the process of criticism; they inevitably exploit and add to past forms of expression, and in exploiting and adding to them, criticize them. It is no coincidence that the greatest artists have often returned to classic, formative works of their tradition for inspiration, works that they realize have been constitutive of their own mode of expression, and to which they want to add their personal comment.

One can, of course, view some artistic work or object in a spirit of scientific detachment, abstracting from any engagement with its human meaning and felt significance. This *may* happen with a first encounter with works from an alien culture, whose forms and language we are unable to interpret as structuring human responses and attitudes, though this is not necessarily the case; sometimes such works, in their freshness, evoke an immediate reaction in us. It can also be done with works from our own culture, and is sometimes done in the name of scientific study, which can be innocent or corrupting of taste depending on the aim and nature of the study. In *A la Recherche du Temps Perdu*, Proust gives poignant expression to a detached 'scientific' approach to the human world. He is full of remorse for cruelly sending his mother away from Venice:

> The town that I saw before me had ceased to be Venice. Its personality, its name seemed to be lying fictions which I no longer had the courage to impress upon its stones. I saw the palaces reduced to their constituent parts, lifeless heaps of marble with nothing to choose between them, and the water as a combination of hydrogen and oxygen, external, blind, anterior and exterior to Venice, unconscious of Doges or of Turner.[18]

The poignancy of the passage arises from the fact that for many, including Proust himself, Venice stands as an exemplar and a symbol of a world transformed by the human aesthetic imagination.

In so far as we are to be at home in our world at all, we need to find our feelings and attitudes corroborated and enhanced in it, and one significant way of doing this is through immersion in those objects which we call works of art, where the world is seen and moulded in terms of human experiences and imagination. Each one of us is, in a way, even in our deepest and most personal feelings a reflection of an already existing human world. Not to understand our cultural heritage and those works of artistic imagination in which its meanings are most explicitly captured and, even more to allow it barbarically to be destroyed is to deprive ourselves and our successors of the sense that we are part of a living tradition of articulation and inculcation of attitude and feeling. And without a sense on the part of at least some members of the tradition of the heritage and what it consists in, it is likely that the tradition itself with its possibilities of depth and nuance of feeling, attitude and experience will die. Certainly no gap that arose here could be filled by science or technology, for science and technology do not deal with what it is to be human at all. (Even if they deal with 'human' subjects, they deal with them in a non-human way.) To return to our Rilke quotation, we have to make ourselves equal to the violin that utterly gives itself, but to do this of any violin Rilke would have been referring to requires an inward understanding of the musical tradition of Bach, Mozart, Beethoven and Brahms. We must also, I think, make ourselves equal to the way the natural world appears to us. But to do this will not be to leave the human world, for one of the most vibrant and recurring features of artistic traditions from Homer onwards is the imaginative response to nature, the attempt to capture and extend in works of art the impact nature makes on us.

We have now briefly delineated two different imaginative approaches to what we are given in experience, one of which eliminates the human perspective altogether as far as it can, and the other which remains tied to human perspective and human response. Even in this preliminary survey, it is clear that it is only within the second perspective that our inner sense of what it is to be alive and in this world, is developed. But now a worrying thought might begin to get a grip. The scientific world view claims to be truly objective. It shows notions such as the vault of heaven to be simply a reflection of a mistaken point of view, an illusion, and from its point of view rather childish. It may be, as we have suggested, that such an illusion cannot be avoided by any creature that exists in space and time, that things will necessarily appear to be like that. It may also be, from our limited perspective, that we cannot altogether avoid taking a human centred view of things, seeing things in terms of colour, and even seeing shadows as scarlet. But perhaps from an objective, scientific, point of view all this too is based on various illusions and partial vision such as our colour perspectives, and that it is from science that we learn what the world is really like. And such knowledge, in any case, inextricably interacts with our reactions to works of art, undermining possibilities of positive response to works – such as Danto's *Purgatorio* and *Paradiso* – which many feel are too firmly anchored in what science has shown to be a false world view. And in learning more what the world is really like may we not, as Wittgenstein suggested, cease to give emotional and moral primacy to our anthropocentric, cultural point of view? May it not, in the end, be better for us to do just that, particularly when so many of the world's ills arise just from men pushing or attempting to push their own perspectives, interests and cultures on others and on the non-human world? Perhaps a scientific realism will not only show our views of the world to be fundamentally limited and partial, but also expose its moral shortcomings and lead us to a better, less destructive approach to existence, to the saintly position mentioned earlier in which one does indeed revalue one's priorities and relationships in favour of a more universal, de-centred approach to life, in which the value of another life on the other side of the earth or even the galaxy really will seem equal to our own, and as much worth cherishing and fostering. Might it not be that the de-centering implied in a scientific realist perspective actually supports better moral conclusions than the stress we have been giving to fostering our own cultural traditions? We need, it seems, to examine more closely than we have so far the claims of modern science to tell us what the world is really like, and to expose much of our common sense perspective as belonging to the egocentric childhood of humanity.

CHAPTER 2

A Decentred Universe: The World of Modern Science

We have already seen some features of the *Weltanschauung* of modern science. In this and the following chapter, I shall focus on this conception in more detail, initially to bring out some of its implications. I will then draw some conclusions about the extent to which science can in fact succeed in attaining a complete and totally objective theory of the world, uncontaminated by human perceptions and concerns.

It is customary to draw a contrast between 'modern' science – that is, science from the seventeenth century onwards – and the work which had gone on in earlier times under the name of natural philosophy, and which was largely inspired by Aristotle. Over-schematic and over-simplified as such contrasts are bound to be, there is in this case at least something to be said for contrasting an ideal-type Aristotelian conception of science with a similarly idealized seventeenth-century conception, because doing this will help to bring out graphically just what is involved in the attempt of modern science to displace man from the centre of the world picture.

The Aristotelian perspective

For Aristotle and the majority of his followers (which would include most scientists down to the sixteenth century), the earth is at rest at the centre of the universe, with the planets and other heavenly bodies revolving round it on a number of concentric spheres. This vision fitted well with the Christian faith, and, as is well known, came under increasing pressure in the sixteenth century leading eventually to the conflicts between the Church and Giordano Bruno and Galileo. From the point of view of the foundation of knowledge, putting the earth in the centre of the universe by itself might appear not terribly important, but it was a significant pointer to Aristotle's methodological views, according to which man was naturally well-adapted to discovering fundamental truths about the physical world. Aristotle believed that through a process of what he called induction, we could move from random observations about the natural world to certain knowledge of it. The first stage of induction is to collect a number of examples of any class of things we are interested in. We might, for example, be interested in giraffes, and notice that all the giraffes we observe have four-chambered stomachs.[1] What we have here is a simple generalisation from sense experience, but the question then arises as to whether the property in question is an essential to being a giraffe. Could something be a giraffe and not have a four-chambered stomach as, presumably, something would still be a giraffe even if it lacked the tufty beard or mottled markings giraffes generally appear to have? According to Aristotle, intellectual insight or intuition will enable the experienced scientist to 'see' whether a property is an essential one or not. In an ideal case, he will be able to expand this insight by showing the reason for the species having the relevant property: that the giraffe, let us say, has a four-chambered stomach in order to deal with the type of food available to it. Ultimately, any fully developed science will be based on first principles which are self-evident or obvious to the scientist, and need no further proof or demonstration, but which explain why certain things behave in the way they do. For example, according to Aristotle, the shape of the universe is 'by necessity' spherical 'for that is the shape most appropriate to its substance, and also by nature primary'. The latter is true because spheres are more simple than other solid figures, while the first body (which the universe is) will possess the first shape.[2]

A number of assumptions underlie all this. First, that what we are given in sense experience is a reliable guide in the matter of the classification of phenomena into natural kinds or species. Secondly, that our minds are structured in such a way that they will naturally intuit the essential aspects of natural phenomena. Thirdly, that what seems a self-evident reason to us for something happening, or what simply seems self-evident in itself is true, and can form the basis of certain knowledge because it could not be otherwise. All these assumptions sit well with the geocentric world view, in which man is at home in the universe, physically and intellectually, but all are questioned in varying degrees in modern science. From the case of the shape of the universe, we can see that what might be obvious to people at one time is very far from obvious to those at another. For us, it is far from clear that spheres are more simple than other shapes or that the first body will possess the first shape (even if we had any grasp of the very Aristotelian notion of heaven as the first body). More fundamentally, even if these things were, in a sense, self-evident to us or intuitively highly plausible, this would not necessarily make them true of the physical world.

The history of science, particularly over the last century, is full of cases of 'self-evident' assumptions being challenged, such as the constancy of physical magnitudes (such as length), or the existence of absolute simultaneity between events. Nor is it an observed fact, or any sort of fact at all that the universe is spherical. That Aristotle believed that it was is perfectly understandable: the vault of heaven *looks* like the inside of a sphere. But claiming that there is such a thing as a spherical body enclosing us is very much a matter of an interpretation of actual observations in the light of prior theory, a phenomenon widespread in the natural sciences, and not in itself objectionable so long as one realises that how one sees and reports observational data is often coloured by one's theoretical preconceptions. More criticizable in Aristotle's case is the prevailing assumption in his thought that natural phenomena are the way they are so that they fulfil purposes our minds tend to impute to nature, as in the case of the giraffe's stomach, or as in his explanation that heavy bodies fall to the earth because that is their natural resting place; that is, that to which they are (in some sense) striving to reach. This form of explanation now seems to us a blatant case of imputing aim or purpose, where there is no purpose; while this criticism is correct, one can also see the imputing of aim more generally as part of Aristotle's natural tendency to see the whole universe in human terms, with things classifiable in terms of their appearances to us and explicable in terms of what seems self-evident to us and in virtue of their having the sorts of motives and purposes human beings have. For Aristotle, man was the measure of all things, and the human mind the mirror of nature, both in terms of its sense experiences and its theorizing.

By contrast with the anthropocentrism and anthropomorphism of Aristotelian science, modern science postulates a world which is fully independent of us or any other observers, in which things might not be at all as they seem to us, and which will not necessarily operate on principles graspable easily and naturally by the human mind. This perspective can actually be grounded in the view evolutionary theory gives us of our sense organs and intellectual faculties. If we and our faculties are adapted to our natural environment, we are adapted to dealing with medium-sized objects in stable environments, such as are found on or near the surface of the earth. There can be no guarantee that we are in a good position to understand or cope with other environments, such as the very small world of microphysics or the very large world of astronomy and cosmology. Radical scepticism of this sort concerning the limits of our knowledge was not to be found in the early stages of the intellectual revolution we call the rise of modern science, but I will go on to suggest in chapter 3 that certain tendencies within that revolution lead quite naturally to it. Here, however, I shall be concerned with the tendency in modern science to present a world view in which our human ways of perceiving the world, through our senses of sight, hearing, touch, taste and smell, are thought to be based on what must be regarded as in some sense illusions founded in our sense organs themselves.

The scientific image versus the manifest image

Galileo had taught that physics was to be restricted to the study of those qualities which undergo systematic quantitative changes relative to some scale of measurement; these so-called primary qualities include shape, size, number, position and 'quantity of motion'.[3] The distinction implied in this doctrine – that only mathematically measurable properties really inhere in things and that other properties are the result of interaction between things and our sense organs – quickly became the basis of seventeenth century physics, and an assumption common to Locke, Descartes and Newton. No more than Galileo were any of these thinkers assailed by any doubts that the secrets of nature would be revealed to those who could read the largely mathematical language of the primary qualities. But the world that was revealed by the new physics was strikingly unlike that of common sense observations or the geocentric universe of Aristotle, in which man and his perceptions and thoughts were comfortably at home.

For Newton, for example, the table before me really consisted of tiny colourless corpuscles which, whether taken singly or in conglomerations, lacked just those qualities of colour, feel, and aural resonance which characterize our actual experience of the table. For Newton, neither objects nor the light rays, which enable us to perceive objects as coloured, are themselves coloured; the rays, which consist of tiny particles moving at great speed, simply have a power or disposition to stir up in us a sensation of this or that colour without themselves being coloured. Sound, as in a bell or musical string is nothing but a trembling motion propagated in the air by the object, perceived by our brain under the form of sound.[4] These views were well known to the seventeenth and eighteenth century philosophers, Locke, Berkeley and Hume, and important in different ways to their respective philosophies.

Locke tended to what I will call scientific realism: that is to say, he claimed that the properties of things which were attributed to them by physics were their real properties and these he called primary qualities.[5] Other properties, such as their colour and taste and sound and texture and smell were not possessed by objects in themselves, but only in so far as the colourless, tasteless, silent, odourless corpuscles which made up those objects affected the sensory organs of those perceivers who were programmed to perceive colour, taste, texture, sound and smell. Berkeley and Hume, as we will see, were quick to seize on and criticize the implication of this view, that we can never perceive the world as it really is, but only as it is filtered through our sensory organs, that we are, in short, victims of a systematic illusion here: our experience is of physical objects having colour, texture, sound, etc., but there are really no such things.[6] Notice that on the Lockean view it is not just that the particles of physics are too small to be perceived. Even if we had microscopes powerful enough to magnify them so they could be seen, and other devices to allow us to know what they felt and sounded and smelt like, we would not be perceiving them as they really were. It would just be that single particles rather than groups of them were now having effects on our sense organs, but we would be no nearer perceiving them as they really were. Conversely, the groups of particles that go to make up the physicist's table lack the sensible qualities of the everyday table, even when they are present in large numbers, as they presumably are in the actual table. What we perceive is always what Wilfrid Sellars calls a manifest image of the thing, rather than the thing itself.[7]

I want to insist that there is a real conflict between the scientific image and the manifest image. According to the world view of the manifest image – the common sense view – we perceive things largely as they really are, allowing for the correction of everyday illusions. That is to say, what we perceive in normal conditions as a blue house we perceive as such because the house itself really is blue. The space in which we live and move is itself structured and made known to us by our apprehension that there is a blue house there, a red car to its right, a brown gate on the left, a bird singing in the green tree behind, and so on. And when we correct an ordinary sense impression, such as that of a blue mountain in the distance, or a brown stick bent in water, we do so by seeing that close to the mountain is really green and brown, or that the brown stick is really straight when taken out of the water. According to the scientific image much of this is mistaken. Our perception of the blue house is due not to the house itself being blue, for it consists of particles that are in themselves without colour or other secondary qualities. Photons – again colourless particles – are emitted from the relevant area of our environment which affect the rods and photocells in our eyes so as mysteriously to create in us a conscious representation of the collection of colourless particles before us as a blue house. (I say mysteriously because brain science can give no account of how the stimulation of cells in our brain produces a conscious experience, replete with secondary qualities which, according to the scientific view are not really in the world at all, but come into existence only through the interaction of stimuli from the world with our sense organs and brains.) In contrast to this scientific story, our ordinary conception of the world would hold that we are conscious of things in the world, and not just of mental representations of them, and conscious of the things in the world more or less as they are, and not as systematically distorted by our sense organs and brains.

According to the scientific world view (since the time of Newton, if not earlier), the blue house I see before me or my son's blond hair are just mental representations of things fundamentally colourless and unobservable except by means of the distortions of sense experience. When Vermeer put his little patch of yellow into his *View of Delft*, the yellowness was not put on the canvas itself; rather than really painting his canvas yellow, Vermeer was actually operating in a world in which there is no colour properly speaking, but which would have an effect on Proust's Bergotte and all the other people who looked at his painting so as to produce in them mental impressions of yellowness.

One way of attempting to remove the conflict between the scientific world view and that of common sense, which was originally due to Locke, is to say that the invisible, tasteless, soundless particles of the scientific world view are to be regarded as the true causes of our perceiving things the way we do. Science leads us to understand that the causes of our perceptions are really the objects of scientific theory, whose existence we infer from scientific analysis of the phenomena. Their existence is inferred because, as we have seen, they are not themselves observable either individually or in the mass.

How, then, are these unobservable causes of my current perceptions identified and located? It might be said that some of the properties attributed by science to objects and the particles making them up are, in fact, perceptible; these would include their position, shape and size, and to this extent the world really is as we perceive it. But, as Berkeley pointed out, our identifications of the positions, shapes and sizes of objects are inextricably linked with our perception of their secondary qualities.[8] You cannot, for example, have a triangle that is no colour at all, or a cube that is not different in feel or colour from its surroundings. In order to pick out an object at all and to perceive its shape or position, then, we are reliant on its having secondary qualities, yet these are just what they are lacking in the case of the supposed causes of the perceptible table, and everything else which occupies and is located in the space in which we exist. But it is through our perception of things having secondary qualities that we actually define the spatial relationships which are at the basis of our actual concept of space. So if secondary qualities go completely, our actual finding of places will go too, and we will be left with the notion of a space none of whose occupants are visible or observable in any other way. The supposed causes of our actual perceptions would then be reduced to being what A. J. Ayer has called unobservable objects in an unobservable space, a surely unattractive proposition.[9]

A possible way out here for the scientific realist would be to say that positions in space are initially picked out by means of our perceptions of objects – with their secondary qualities and all – but we then displace the ordinarily observed objects with the unobservable particles and forces that are said to cause our perceptions. These unobservable things will be said to exist just where we erroneously thought that the observed things existed. Such a proposal, though, should give cause for concern to the scientific realist because his whole story regarding the existence and location of what really exists is now crucially dependent on his initially taking as acceptable the actual perceptions we have of objects which he is just about to dismiss. It seems that science needs our ordinary world of experience at the very moment it is about to reject it.

In addition to relying on the common sense world view to locate objects, the scientific realist will also have to give some account of his reasons for thinking that his theories about unobservable objects and the rest are actually true, and actually describe the true nature of the world. Locke's own defence of the primacy of primary qualities was singularly unpersuasive, relying largely on the repeated assertion that primary qualities are inseparable from the object to which they belong, to which the rejoinder that without secondary qualities neither primary qualities nor objects can strike us at all seems refutation enough. In other words, secondary qualities too are inseparable from our conceptions of objects. (Significantly perhaps, as pointing to the hidden agenda of his philosophy, Locke mentions the measurability of primary qualities as a further reason for their primacy over secondary qualities; between the latter he thinks only qualitative differences can be perceived.[10])

A more plausible tack for the scientific realist might be to appeal to the success his theories have in accounting for the phenomena, but this, too exposes him to an immediate difficulty. His theories assert the illusoriness of much of the common sense world, but our reasons for accepting them are that they produce testable and verifiable predictions in the everyday world. Once again, the attempt to displace our pre-theoretical grasp of the everyday world in favour of mathematically rigorous theory depends for its acceptability on the deliverances of everyday experience. The scientific realist may be in danger here of removing the very ground from which his theories gain support, for as Strawson has pointed out, it is only because we are subject to the illusions of common sense that we are able to engage in the scientific activity of confirming and testing theories at all.[11] Some would be tempted to conclude at this point that scientific theory should be treated not as the only or the primary truth about the world, but rather as a means of providing mathematical models of the functioning of aspects of the world, in which, for example, colour and sound are treated as correlates of wavelength, rather than as a 'false imaginary glare' or rattle somehow eking out a questionable existence on the back of really existing wavelengths.

Perhaps rather than thinking of the things of which science treats as the merely inferred causes of our perceptions in a merely inferred space (before dismissing the experienced world as merely illusory), the scientific realist might prefer to take the world of experience at face value, and to think of the colourless, soundless, textureless corpuscles which science postulates as being the minute parts which *en masse* compose the coloured, resonant and textured tables and houses of our experience. On this view, the basic particles would be regarded as too small to be observed, but as taking on observable properties when grouped in masses. The objection to this rather natural way of thinking is that it is simply inconsistent with the scientific image. According to Eddington's famous discussion of the 'two' tables, the table of common sense on which the author wrote and its scientific counterpart, the 'scientific' table consists largely of empty space, occupied intermittently by groups of rapidly moving, colourless (etc.) particles and mysterious forces which bind and separate the particles in various relationships of repulsion and attraction.[12] Even from the scientific point of view we cannot *see* the real scientific table as it 'really' is; all we can *see* is the illusory image of the brown, smooth, stable, polished thing I am writing on. And Eddington's position is surely correct on this point. Science does not teach that groups of elementary particles have observable properties, even when the groups are large, except in so far as they create in our brains observable images of the unobservable things.

Similarly, from the point of view of the manifest image of common sense, parts of something observable, however small, remain in principle observable. The transition to the unobservable particles of science from the everyday objects we observe cannot be regarded as like seeing the true colour of, say, blood (or the cells that make up blood) under a microscope. Even though red blood appears pellucid under a microscope, we are not then seeing how blood really is, scientifically speaking, for a pellucid fluid may not be red, but it is still coloured.[13] The two world views, then, cannot be combined in any straightforward way: one teaches that the brown polished table top is in itself nothing but unobservable particles in an unobservable space, however many particles and however large the space you take; the other that observable properties inhere in the things themselves, as far down as you like to go, and are not just figments of the way we represent things to ourselves. Indeed, ordinary realism holds that we perceive the things themselves, and not images somehow caused by them. In giving a scientific description of the way the table is in itself, on the other hand, you just would not mention its colour or texture, any more than an account of the physics of the canvas and paint that make up the *Mona Lisa* would say anything about the smile or the face. The smile and the face and the colour and the texture of the table are all, from the scientific point of view, simply features of the way that certain collections of particles in particular spaces appear to human beings.

There is another way of attempting to resolve the tension between the scientific approach to reality and the common sense attitude, which involves neither taking the particles postulated by science to cause our manifest image of the world nor thinking of them as the ultimate constituents of the ordinary things we observe; it is to say that the same thing – Eddington's table, let us say – can be looked at in two ways, *either* as the object of everyday perception *or* as the collection of imperceptible particles postulated by physics, depending on our particular interest at the relevant time. Thus, for example, Mary Hesse speaks of the scientific world view and the everyday one as simply different ways of describing the same things,[14] much as if I were to describe the table-top before me as either a table-top or a piece of rectangular wood. But what is important here is that unless we take the scientific description as merely a phonetic or typographical variant of the everyday one – so that some speak of 'exerting repulsive force' where most people would speak of 'hardness' – in the scientific description, we are talking about things in terms in which they cannot be observed at all. Strictly speaking, there is no scientific *point of view*; science provides a conception of a world lacking observable properties, and also of things having strictly unobservable properties, such as momenta without position, and in which observable properties get somehow introduced only in the interactions things have with us and other sentient beings. The problem this conception raises is that our contact with the world is through the observable properties of things. Our common-sense conception of the world is of it being filled with objects with observable properties. Treating the scientific and ordinary conceptions of the world as simply two ways of looking at the same thing raises in an acute way the problem of explaining just what thing is being looked at in two different ways. To repeat, it is not like treating the object I am writing on as either a table-top or a rectangular piece of wood. There is no contradiction in saying that a table-top is a piece of wood, but there surely is a contradiction in saying that the same space is occupied by something both coloured and colourless. What type of thing is it that can have both properties?

The explanatory power of the scientific picture

The answer the scientific realist will give to this question is to say that strictly speaking and in themselves objects do not have observable properties, and that these arise only in the images or representations we have, which emerge from the interactions that exist between us and the world. Our common-sense picture of the world is one centred on us and our perceptual faculties and which (misleadingly) attributes to things properties they do not have in themselves. We must de-centre our world-view. Just as the earth is not the centre of the universe, neither is our original conception of the world what it seems to be: a direct and broadly correct impression of how things in the world are in themselves. The scientific realist would admit that we confirm the theories of science through their power to predict observable effects in our common-sense world: as the physicist Richard Feynman puts it, you can find out whether the consequences of a scientific theory are true only if you can translate the theory's claims into 'the blocks of copper and glass that you are going to do the experiments with.'[15] Moreover, the scientist will depend on common-sense perception initially to identify the objects he is talking about and their spatial positions. But, it will be said, science is entitled to claim that its world-view is really more fundamental than the common-sense one because it *explains* how it is we come to have our erroneous common-sense view of the world as populated with objects having humanly observable properties. The explanation will be given in terms of its conception of the world as consisting of fundamentally unobservable entities becoming observable for us in interaction with our sense organs and brains. In this interaction the colourless photons which hit our retinal surfaces initiate a process which culminates in the construction in our brains of representations of a world populated with objects having colour and the other observable properties.

The reason why science can claim to tell us what the world is really like, absolutely speaking, is now clear. It derives from the claim of science to present us with theories which are not only true, but which at the same time correct and explain (and correct *through* explaining) our common-sense perceptions of the world. Leaving aside questions as to the truth of scientific theory, how plausible, though, is the view that science can explain our common-sense perceptions of the world?

The fundamental difficulty here is a simple one, and actually goes back to Locke himself: it is to explain 'how any size, figure, or motion of any Particles can possibly produce in us the Idea of any Colour, Taste, or Sound whatsoever; there is no conceivable connexion betwixt the one and the other.'[16] I do not believe that any progress has been made on this problem since the time of Locke. While mathematical correlates of colour vision, hearing and the rest may have been discovered, what has not been explained is how streams of photons interacting with the eye or other particles interacting with the other sense organs can produce the conscious experience of colour. For our visual experience is not like the operation of a camera, nor do we function exactly like robots or computers with sensory apparatuses attached. One can perfectly well envisage a machine responding to and reacting to environmental stimuli; but all that can be explained in terms of its responding to stimuli which triggered off bits of its programme. There would not be the slightest reason to assert that the machine had any sensations of which it was conscious, or that, like Bergotte, it was capable of being suffused with the experience of the precise quality of Vermeer's yellow. But it is with just such experiences and ideas that our mental lives are filled. The difficulty is to understand how neural activity produces conscious experience. And once again, we must not be misled by all too easy talk of representations in the brain. If a voltmeter or camera or computer 'represents' states in the outside world, then our conscious representations are quite different; if, on the other hand, our representations are, neurologically speaking, bits of electro-chemical stimulation in various points in the cortex, then what needs to be explained is how we are conscious of them as yellow patches, blue houses and the like. Two further familiar problems with talk of consciousness in terms of representations in the brain are the so-called 'homunculus' problem about who or what it is inside the brain that is perceiving the representation, and the old Lockean problem of knowing how far a representation in the brain is an adequate picture of the external world. But we need not delay on these problems unless the initial difficulty of explaining how physical processes can create conscious states can be overcome.

The theorising bat

This initial difficulty is brought out rather engagingly (and unintentionally, I should add) in Richard Dawkins's *The Blind Watchmaker*.[17] Dawkins attempts to tackle the question raised by Thomas Nagel in his article 'What is it like to be a bat?',[18] Nagel's point being that it must be like *something* to be a bat, but something on which from our sensory standpoint we might not be able to say anything, given what would appear to be the considerable distance the bat's experience is from our own. Dawkins quite correctly observes that being a bat would not be like you or me going into a cave, shouting or banging two spoons together, consciously calculating the delay before hearing the echo, and calculating from this how far the wall must be from us. This would be no more a description of the bat's subjective experience than our experiencing blue or red is to be analysed in terms of our calculating from the wavelength of the light entering our eye whether the wavelength is short or long and concluding from that that we are perceiving red (long wavelengths) or blue (short wavelengths). Dawkins concludes that though the bat perceives the position of an insect by using what we call echoes, it 'surely no more thinks in terms of delays of echoes when it perceives an insect, than we think in terms of wavelengths when we perceive blue or red'.

All this is a sound reminder of the difference between talk of the nature of conscious experience and its physical basis, but Dawkins then goes on to speak of a percept possessed by a conscious being of something in its environment as an 'elaborate computer model in the brain', in which wavelength differences are coded as colour differences, shape differences, and so on. These differences, and the subjective differences between sight and sound and the rest are properties of the internal model, and not of physical differences between light or sound or the physical attributes of nerve impulses, which, he says, are the same for all sensory modalities. What happens is that outside information and nerve impulses are translated into the computer models, which are then in turn subjectively experienced as visual fields, aural fields, and so on. This being so, according to Dawkins, a bat's experience, given its needs in the world, could, though stimulated by ultrasound rather than the visual part of the wavelength field, be much the same as our visual experience:

> I don't mean this just as some vague metaphor. It is possible that the subjective sensation experienced by a female bat when she perceives a male really is, say, bright red: the same sensation as I experience when I see a flamingo.

Dawkins envisages a scientifically-minded bat commenting on our visual experience, and saying that we humans use inaudible rays called 'light' in conjunction with specialized organs called 'eyes' to exploit the complex echoes that bounce off objects when light rays hit them, in such a way that there are one-to-one mappings between objects in the world and 'images' on sheets of cells called the 'retina': 'These retinal cells are capable, in some mysterious way, of rendering the light "audible" (one might say), and they relay their information to the brain.'

Dawkins's fantasy is charming, but he gives everything away in the second phrase of the last quoted sentence: 'in some mysterious way'. Even before we get to the problems of the homunculus being conscious of the retinal image or of the comparisons of the retinal image with reality, we are told that for the theorising bat it is mysterious how for us light becomes 'audible'. Is it any the less mysterious to us to see how for us light becomes 'visible', meaning by that giving us subjective experiences of seeing colours, shapes and so on? It would not be unfair to say that neither Dawkins nor anyone else has provided even the form of an explanation of this question. We are all reduced at this point to talking about mysteries.

The scientific realist will no doubt reply to this that the common-sense view of the world as consisting of objects as having observable properties does not provide an explanation of human experience either. To this, the defender of the non-illusory nature of common sense could reply that at least it explains why we see things as coloured, hear them, feel them, and so on, because colours, sounds, tactile properties and the rest really exist in the world. But I do not think it really matters whether or not the common-sense view has more explanatory power than the scientific view: one could quite fairly point out that the common-sense view is not in the business of offering *explanations* of sensory processes. Far more decisive, though, against the claims of science to offer an account of how the world really is is the fact that its claim to be able to correct the common-sense view of the world actually rests on the further claim that it can explain how that view arises (and hence, why and in what respects it is illusory). What we have just concluded, however, is that its explanation of our having sensations at all is quite empty, amounting to no more than correlations of, say, visual experiences with wavelengths. Experience itself remains totally unexplained, for much the reason given by Locke. And this should make us sceptical of any move to assert the primacy of the scientific world view over the common-sense view on the grounds that the former can explain, and by explaining correct the latter.

We have, I think, found in this chapter a strong reason for being sceptical of the claims of science to present a more complete and objective account of how the world is than our ordinary vision of the world, given that it is quite unable in its own terms to explain how we come to have that vision of a world filled with secondary properties and subjective experiences. It is, in other words, unable to explain how creatures like us come to be putting forward scientific theories at all, given that, as we have seen, doing this depends on our secondary quality perception – and this must surely undermine any claims made on behalf of science to present a complete account of how things are. At the same time, we have elaborated in some detail the contrast between the view of the world of common sense and what I have been calling the decentred account of the world presented by modern science. The question that we now have to consider is whether, despite its inability to account for human experience, there are any other grounds for regarding the theories of modern science as true, at least in so far as they do not touch on subjective experience. Or are there further problems, not yet touched on, if we want to take the theories of science at face value?

CHAPTER 3

Science and Truth

In *The Gay Science*, Nietzsche wrote:

> Do you really believe that the sciences would ever have originated and grown if the way had not been prepared by magicians, alchemists, astrologers and witches whose promises and pretensions had first to create a thirst, a hunger, a taste for *hidden* and *forbidden* powers? Indeed, infinitely more had to be *promised* than could ever be fulfilled in order that anything at all might be fulfilled in the realm of knowledge.[1]

As so often, without benefit of supporting evidence or arguments, Nietzsche gets to the heart of the matter. Thanks particularly to the researches of historians such as Frances Yates, we now know far more than Nietzsche could of the mystagogical origins of modern science and, of the significance of an esoteric approach to nature, deriving from Hermetic and Cabalist sources in the development of a mathematical approach to natural phenomena. In the seventeenth century, side by side with what has now come to seem the 'official' empiricism of patient and systematic observation of nature, went interest in activities such as alchemy, astrology and numerology, without necessarily provoking any discomfort. Newton himself, as is now well known, dabbled extensively in alchemy, perhaps in the hope that something there might be revealed that his physics could not see.

Certainly, no slavish empiricist fact collecting could have been the basis of laws as simple and general as Kepler's or Newton's which so far transcend (and even occasionally fly in the face of) observation. Indeed, pure observation was not the inspiration: Kepler was motivated by a Pythagorean faith in the fundamentally mathematical nature of reality, in the idea that beneath the appearances, the behaviour of the planets must manifest mathematical harmony, and it was this that led him to search for mathematical formulae which would account for the actual behaviour of objects in the solar system, while Newton saw his mathematization of the physical world as part of a general theological scheme, encompassing and uniting the earth, the heavens and the spiritual realm, and revealing the divine Unity in nature.

Nowadays, of course, of the discoveries and researches of figures like Kepler and Newton we take seriously only the physical ones, but even here deep questions can be asked about the nature of their scientific activity. A Pythagorean, like Kepler, would insist that if he managed to discover a mathematical formula such as his second law of planetary motion, which fitted (more or less) the observed facts, then something deep about the nature of reality had been discovered. Similarly, Galileo insisted that the Copernican system which placed the sun rather than the earth at the centre of the universe was not just a better way of systematizing the data than a geocentric system. It was a true description of reality, and it was this insistence, of course, which led him into conflict with the Church. The Church though was not as obscurantist as has often been projected: Cardinal Bellarmine was perfectly happy to concede that the Copernican model was a better instrument for calculation than the Ptolemaic, so long as one did not infer from that that the Copernican theory gave us true insight into the nature of reality.[2] And this position concerning the relationship of scientific theory to reality, is not, as we shall see, either obscurantist or even particularly implausible when we consider theories which purport to deal with deep levels of reality, far beyond our powers to penetrate by direct observation. (Is it completely clear, even today, that the decision between a Copernican or a geocentric system can be made entirely on *observational* grounds, even though the entities spoken about in the two systems are all observable?)

Newtonian questions

A further initial question about the status of the theories of seventeenth century science comes out in considering Newton's three fundamental laws of motion. In Newton's own words, these state: (1) every body continues in its state of rest or of uniform motion in a right (i.e. straight) line unless it is compelled to change that state by forces impressed upon it; (2) the change of motion of a body is proportional to the motive force impressed, and is made in the direction of the right line in which that force is impressed; and (3) to every action there is always opposed an equal reaction, or, the mutual actions of two bodies upon each other are always equal, and directed to contrary parts. But his law of gravity states that the attractive force between two bodies is proportional to the product of their masses divided by the square of the distance between them. And this in turn means that no object anywhere in the universe can ever be free of gravitational forces impressed on it by other bodies. So no object in the universe can ever be in the condition specified in the first law (under the influence of *no* impressed force). Is the First Law, then, literally true? If it is, it expresses a truth we could never check by observation. Or is it, rather, an idealized abstraction from the motions of things we do observe – a theoretical limiting case useful for making calculations about actual bodies, behaving under actual impressed forces?[3] In similar vein, we can ask whether, on Newton's principles, all bodies have mass. That they do would appear to be suggested by Newton's conception of motion or 'momentum', which, according to his own Definition II is mass multiplied by velocity. This definition is the basis of the more familiar formulation of the Second Law (due to Euler in 1749), according to which force equals mass multiplied by acceleration (f = ma), which would most naturally seem to envisage that we think of bodies having a constant mass in whatever state of motion they are in. And this, in turn, would be a deep and pervasive truth about bodies generally. But, as far as predicting and systematising the observable world goes, we could just as well assume that bodies which never collide with other bodies, or which are never accelerated, either have no mass at all, or have some arbitrarily given mass. Newtonian theory can be formulated on these somewhat perverse-seeming assumptions and yield identical predictions to those yielded on the more natural assumption that non-accelerated and non-colliding bodies have the same mass as they would were they to be accelerated and collide.[4]

What these two examples suggest is that the literal truth or falsity of the First Law and the precise interpretation of Newtonian mass cannot be fixed by comparing the Newtonian laws and concepts with reality. In a significant sense, reality does not determine such things, and this, in turn, raises the question as to the status of the laws and concepts themselves: are they to be regarded as giving literally adequate or inadequate descriptions of reality, or are they to be regarded as tools or idealizations or abstractions, more or less useful for calculating, predicting and dealing with natural phenomena? On the latter view, which I shall call the non-realist interpretation of scientific theory, we could give good reasons for operating with the First Law or with one particular assumption about mass (say, that all bodies have all the time the mass they would if they collided with something else or were accelerated). On the non-realist view, these good reasons would be in terms of the simplicity, applicability, generality of the theory of assumption and in terms of its coherence or fit with other theories we hold, but we would not regard any of these considerations as ensuring that the assumption or theory we chose was literally true.

I have dwelt on examples from the early period in modern science, because I want to emphasize that even then philosophical questions could be legitimately raised about the status of theories which appeared to inform us about things which could not be checked by observation. It is almost a platitude now to say that Newtonian physics had become so successful by the eighteenth century that it needed an effort of will, almost of perversity, to question its title to be the absolute, literal truth about the world. Some philosophers, such as Berkeley and Hume did make the effort, expressing doubts about the real existence of the unobservable forces and entities Newtonians postulated. In this, curiously, they were partially anticipated by Newton himself at least as far as unobservable forces went. He was never happy with the thought that gravity could be an ultimate constituent of the universe, or that action at a distance was something to rest satisfied with. In the General Scholium to Book III of his *Mathematical Principles* he says that there undoubtedly are the effects his law describes: that the planets and the sea are moved by the power of gravity, and that the force of gravity depends on the mass of the bodies involved and propagates its virtue on all sides to immense distances, decreasing always as the inverse square of the distance, but

> hitherto I have not been able to discover the cause of those properties of gravity from phenomena, and I frame no hypotheses; for whatever is not deduced from the phenomena is to be called an hypothesis; and hypotheses, whether metaphysical

> or physical, whether of occult qualities or mechanical, have no place in experimental philosophy.

What Newton says here is doubly strange; first, because an obvious reaction to his remarks is to say that gravity itself is the cause of the phenomena in question, and secondly because Newtonian physics is generally regarded as the greatest example of a system which, in the words of Pierre Duhem explains by stripping 'reality of the appearances covering it like a veil, in order to see the bare reality itself'.[5] But even if his *Opticks* may be said to be doing just this on light and colour, apparently Newton himself was more cautious about the pretensions of his work when it came to an occult force like gravity. And, it might be observed, he had good reason to be cautious at the latter point, because Einstein accounts for the same gravitational phenomena as Newton without postulating a gravitational force.

But this is to anticipate a little. My aim so far has been to suggest that even within the Newtonian system, there are grounds for questioning whether scientific theories are to be taken as literally true (or false) descriptions of reality. In so far as they speak about what cannot be observed, there are grounds for treating them non-realistically, as useful instruments or models for organizing data, in idealisations, abstractions, and the rest. Despite Galileo, Bellarmine's philosophy of science was never a non-starter, although the success of Newton's physics convinced many that Newtonian theory just had to be literally true, in all its detail and all its claims. How else, it was urged, could one explain its success in predicting all the phenomena if what it said about the hidden realities underlying the phenomena was not also true?

This greatly worried Kant. Unlike Aristotle, he did not believe that one could simply assume that the human mind reflected reality. So how was it that the human mind could come up with a theory like Newton's that was apparently so adequate to the way the world was? The probability of its doing this by chance were practically nil. On the other hand, what sort of explanation, other than a theological one, could there be of the mind's attunement to nature? Kant took the heroic step of explaining the mind's attunement to reality – as supremely manifested in Newtonian physics – by saying that there is a sense in which we construct the observable world, that what we observe and its organisation is partially due to the construction of our minds. We perceive things as related temporally and spatially, as causally related and as enduring substances because this is the way our minds are formed to perceive and understand the world. How the world is in itself, apart from the way it appears to us, we cannot say. But we can frame theories about what we observe which are pretty accurate at that level precisely because the level of observation is something we have a hand in constructing.

What from our point of view is interesting about the Kantian position is that it can be seen as in part a response to the very success of Newtonian physics. Newtonian physics apparently gives us deep truths about the observable world not because our minds are in an Aristotelian way attuned to the observable world, but for just the opposite reason – because, in effect, the observable world, just because it *is* observable, is attuned to our mind. So even at and actually because of the height of confidence in Newtonian physics, it was possible for a deep thinker to reject both a totally realistic interpretation of its theories, as telling us about the world as it was in itself, apart from our observations of it, and the argument that the theories must be true because they best explain the world as we observe it.

The inference to best explanation

I shall follow normal practice in calling the idea that a theory must be literally true even at the level of what cannot be directly observed *if* it explains what can be observed better than any other theory the 'inference to best explanation'. This inference must have seemed highly tempting during the heyday of the optimism that Newton and his followers had indeed revealed the secrets of the universe. The predictions of Newtonian physics had been confirmed in countless observations and experiments. All sorts of new discoveries, from the existence of hitherto unknown planets to formulae covering the motions of pendula, from investigations into light rays and the development of atomic theory to Faraday's researches into electromagnetic phenomena had been made possible by the Newtonian framework. The notion that the universe itself was basically as Newton had conceived it, and that any unsolved problems in physics and astronomy were simply waiting for their Newtonian solution, must have been very tempting to scientists in the eighteenth and nineteenth centuries. But, as everyone knows, it just did not happen. The overthrow of Newtonian physics by relativity and quantum theory involved the dropping of many of Newton's most central assumptions, from absolute space and time and determinism to the constancy of physical magnitude and (for many) the very belief that the act of observing phenomena does not significantly interfere with them. What price now a Newtonian inference to best explanation?

Someone who knows this better than most is Sir Karl Popper; for, as he records in many of his writings, it was the overthrow of the Newtonian framework by Einsteinian relativity that led him to see the full significance of the asymmetry between confirmation and falsification in scientific theory. However many times and however triumphantly a theory – such as Newton's – is confirmed by its predictions being observed to be true, its confirmation avails it not if and when it is falsified by one of its predictions turning out false. And if we can learn anything from the past history of science, it is surely that no theory will survive for ever. In view of all this, which Popper has done as much as anyone to emphasize and establish, it is, to say the least, surprising to find Popper himself apparently using a version of the inference to best explanation in connection with Einstein's theory: 'we can argue that it would be a highly improbable coincidence if a theory like Einstein's could correctly predict very precise measurements not predicted by its predecessors unless there is "some truth" in it.'[6] It is true that he goes on to say that this probability of 'some truth' does not mean that it is probable that the theory is true *simpliciter*, but only that it is probable that it has both a high truth content and is nearer the truth than those of its competitors which made less good predictions. But one cannot avoid concluding that if this passage means anything at all, then it must mean that Einstein's theory probably gets nearer to the underlying hidden essence of the world than Newton's theory did. If it means that, though, then it is wrong, and wrong not only because the inference to best explanation is subject to historical counter-examples, such as the demise of Newton's theory itself, which means that it can no longer be taken as an account of the world's essence, despite its tremendous predictive success both absolutely speaking and relative to its predecessors.

Logically speaking, the situation with a scientific theory is as follows. A scientific theory is a statement or set of statements which enable us to make predictions about the observable world. Some theories may stop at this level, and simply assert correlations between observable events, such as a theory expressing by means of a mathematical formula some observed or observable relationship between phenomena. An example here might be Kepler's second law, which states that the radius vector from the sun to a planet in orbit round the sun sweeps over equal areas in equal times. But other, in a way more interesting, theories do more than this, by attempting to strip reality bare, in Duhem's phrase, and to go below the observable surface to the inner essence of things, to what explains observed correlations and causal relationships. It does not matter for the moment just what we count as observable, and how far we regard sophisticated instruments, such as electron microscopes, as extending our powers of observation. Even if, implausibly in the view of some, we count some sub-atomic particles as part of the observable world, there may still be elements in a theory which treat things which are, in a fundamental sense unobservable, such as a force like gravity, or the curvature of space, or absolute space and time, or fields of force, or alternative possible states of affairs. Of all these things, like most elementary particles even on the most generous interpretation of the operations of cloud chambers and particle accelerators, it is true to say that we can at most observe the effects. And unlike Man Friday's footsteps, observation of effects in these cases can never be complemented by direct observation of the cause. The question we now have to ask is: logically, how strong is the inference from effect to cause when the postulated cause is something we can never envisage ourselves having direct observation of?

And the answer is: not very strong at all, given that theories presenting incompatible accounts of how the world is at the deep level can all agree at the observational level and make very precise and very similar predictions at that level. The inference to best explanation would have to hold in each case that it was very strange if success of this sort did not reflect truth at the deeper level. But then, the inference to best explanation would be 'proving' the truth of incompatible deep-level theories. And that is unacceptable. If this type of case seems impossible, consider the following simple example, due to Hilary Putnam. Imagine a world consisting simply of straight lines. Now we know that the geometry of straight lines, which tells us all we need to know about lines and their relationships, can be constructed in different ways, either by taking points to be the fundamental elements of lines, or by regarding points as abstractions from intersecting line segments. And we can imagine two straight line physicists, each hammering away with the inference to best explanation from his observationally entirely adequate geometry, one proving that there are and the other proving that there are not 'really' points.[7]

What all this shows is that wherever there is a level of theory which goes beyond your data base, you move on to shaky ground, because there are, in principle, other incompatible theories which could also explain and entail the same data.There do not, of course, have actually to be any such theories. It is just their possibility that undermines the inference to best explanation. We may forget this when doing science, because at any given time there is usually a marked absence of incompatible explanations and a general consensus about the basic structure of the universe. But if you look at science historically, the opposite turns out to be the case, and fundamental explanations, and pictures of what the world is like come and go, and sometimes even come back again. This aspect of the scientific search for hidden and forbidden powers is something we, post-Newton, post-Einstein and post-Thomas Kuhn's *Structure of Scientific Revolutions*[8] are rather more aware of than Nietzsche could have been, but once more Nietzsche's fundamental insight appears to be vindicated.

Rather than going into any of Kuhn's grand examples of theoretical frameworks not surviving scientific revolutions, let us focus on a smaller example of a change of underlying theoretical structure, which is discussed by John Watkins.[9] Watkins' example brings out rather well some significant features of a transition from one underlying theory to another. It concerns the transition from the caloric theory to the kinetic theory of heat. According to the caloric theory, 'the reality behind phenomena involving freezing, melting, vaporisation and changes in temperature, is movements of a heat substance distinct from ponderable matter', as Watkins puts it. This heat stuff obeys a conservation law, but it is always tending to disperse itself, so when a hot body meets a cold body, the caloric flows out of the hot body into the cold until both reach the same temperature. The kinetic theory, by contrast, denies that there is any such stuff as heat. 'What is conserved through all heat changes is not caloric but *energy* ... behind heat phenomena there are swarms of tiny molecules moving randomly, often with enormous velocities, and collectively possessing a mean kinetic energy.' More heat means more molecular movement. When a hot and a cold body meet, the molecules near the surface of the cool body move faster and collide more frequently with those further in. Thus, 'the mean kinetic energy of its molecules is raised and that of the hotter body's is lowered, until a uniform level is reached. *So it is as if heat flowed out of one into the other*' (my italics).

Why this example is so pertinent is because it shows clearly how underlying ontologies or theoretical structures work in science. What they do, in effect, is to postulate 'as if' stories, showing that it is possible to regard a given phenomenon – in this case heat – *as if* it was brought about by 'an elastic fluid of great subtility' (Dalton's description of the caloric he postulated), or *as if* it was brought about by molecular motions or bombardments. The kinetic theory superseded the caloric theory, but at an observational level the two theories had to respect the same phenomenal or observational laws relating to the behaviour of heat phenomena. If the kinetic theory had not been able to predict the known heat phenomena more or less as well as the caloric theory, it would not have been taken seriously. At the same time, while denying the underlying ontology of the caloric theory in the most fundamental way possible, by claiming that there is no such thing as caloric, the kinetic theory explains how people could have come to think of the movement of caloric as the cause of heat phenomena. The changes of movement of molecules could look like the flowing of caloric, and so, as Watkins puts it, 'the phenomenal world behaves *almost as if* the reality behind it *were* as (the caloric theory) declares it to be.'

Observation without theory

It is plausible to suppose that many if not most scientific revolutions are somewhat like our heat example. Followers of Kuhn might object to this claim on the grounds that a change of scientific theory will often entail a change of observational assumptions, too, so that there is no common world of observation explained by each of two (or more) competing theories. While it is true, though, that one's ideas or theories about what it is one is observing may well be theory-laden and affected by one's theoretical assumptions, it does not follow that at some level people from different theoretical backgrounds cannot recognise that they are both talking about and trying to explain the same phenomena. Kuhn himself takes as an example of theory affecting observation the case of the pendulum as interpreted by the Aristotelians and then by Galileo and his successors. Kuhn writes:

> since remote antiquity most people have seen one or another heavy body swinging back or forth on a string or chain until it finally comes to rest. To the Aristotelians, who believed that a heavy body is moved by its own nature from a higher position to a state of natural rest at a lower one, the swinging body was simply falling with difficulty. Constrained by the chain, it could achieve rest at its low point only after a tortuous motion and a considerable time. Galileo, on the other hand, looking at the swinging body, saw a pendulum, a body that almost succeeded in repeating the same motion over and over again *ad infinitum.*[10]

Kuhn's point is that the Aristotelians saw the phenomenon as an example of constrained fall, while Galileo and we see it as a pendulum. These are the terms under which we perceive it, terms which are in part determined by the respective theories we each hold about the thing. But it does not follow from what is true in all of this that we and the Aristotelians cannot agree about what it is that we are each categorizing and even perceiving in our own ways. Kuhn himself, in speaking of a heavy body swinging back and forth on a chain, provides a more basic and less theoretically contentious form of description which both parties could presumably agree was a basic account of what they were talking about.

But even where a less disputed and mutually acceptable form of description may not be easily available, it is simply not the case that people theoretically at odds with each other cannot come to realise that through their theoretically charged and mutually contradictory terms and descriptions they are actually referring to the same things. Thus, Joseph Priestley thought that when he poured sulphuric acid on zinc he observed phlogiston being emitted. Phlogiston was believed by eighteenth century chemists to be an indirectly observable gaseous fluid which was flammable, had no weight or even negative weight, produced pure metals from rust, and was emitted when sulphuric acid was poured on to zinc. We now believe that there is no such thing as phlogiston, meaning by that that there is no thing that has all the properties listed. We say that *hydrogen* is emitted when sulphuric acid is poured on zinc; but, as Mary Hesse argues, 'there is no doubt that *in these* (sulphuric acid-zinc) *experiments* Priestley recognized and referred to "hydrogen" – he just called it "phlogiston".'[11] Although Priestley was wrong in many of his beliefs about phlogiston, we, from our point of view can say that 'Priestley knew many truths about "hydrogen"'; and presumably Priestley could have said of us that we knew many truths about 'phlogiston' (the ones relating to acids and metals). The point is that both parties to the dispute could recognize that they were talking about the same substance when talking about the gas (phlogiston/hydrogen) that was emitted when acid was poured on a metal. We can also both recognize many low-level truths about such experiments and about the gas emitted, and these observationally based truths, like the heat phenomena considered earlier and the behaviour of swinging stones, will form part of the growing body of knowledge which acceptable scientific theories will have to take into account. This type of knowledge is what accumulates in science, while there is now a large consensus among philosophers and historians of science that there is far less continuity and convergence at the level of 'deep' scientific theory or theory about the unobservable structure of reality. At that level – the level where observation does not control theory – the history of science looks altogether less smooth, and science itself looks much less assured by hard data, and far more a function of human imagination and conjecture.

Not all the examples we have been considering are of cases where the underlying theory is totally inaccessible to direct observation or evidence. In the heat case, for example, molecules and their movements would now be widely regarded as matters of fact and observation. There can, of course, be cases where the imposition of a theory itself leads to the construction of observational instruments and techniques, and a corresponding growth in the hard, factual core of scientific knowledge. But equally, as I have already suggested, much scientific theorizing goes on at levels quite underdetermined by data, and change from one theory to another is, as Kuhn has argued persuasively, much more like a switch of *Gestalt* or a religious conversion, in which it is not so much the world that changes as that we now look at the world in a different way. We stop, for example, seeing the world in terms of things striving or tending to reach their naturally appointed goals (as the Aristotelian viewpoint would have it), and see it simply in terms of billiard balls knocking each other about and cogs and wheels interacting (the paradigm adopted in the early eighteenth century). Do the facts determine choice of theory at this sort of level? Do the facts say that the Aristotelian way is definitely wrong? What we can say is that the mechanist billiard ball and cogs and ratchets view of the universe has not prevailed, and was, of course, dealt a severe blow by Newton's introduction of a gravitational force which postulates action over great distances. It is interesting to note as we have seen that Newton himself was sufficiently influenced by mechanist ideas to be quite ambivalent towards his concept of gravity and its explanatory power, which was regarded by many as a throw-back to the discredited medieval type of explanation in terms of occult qualities. But the mechanist is still free to say that underlying gravity, there are undetectable little particles knocking into each other and even theorise about what these things must be like to produce the gravitational effects we observe. And has even Aristotelian talk of tendencies in things and their natural states gone forever, when contemporary physicists seem prepared to envisage states of affairs with innate propensities to decay and to attract and repel other particles?

What I am now referring to is the ground that has been well traversed by Kuhn, the phenomenon by which deep explanatory conceptions in science – models of what the universe is actually like at the deepest level – have come and gone, and sometimes come back again. Against Kuhn, I have been suggesting that there is a degree of consistency and cumulative growth in science at the level of observation, and that this can be recognised even where the observations themselves and the accounts of them are highly theory-dependent. If this is true, though, it simply emphasizes the extent to which fundamental theory floats free of observation and experiment, perhaps drawing investigators on to new levels of observation and new types of experiment, but never entirely pulled down to earth by the facts as we are able to grasp them in observation. Theory asks us to conceive the unobservable world as if it were a system of cogs and pulleys, or forces and fields, or particles and anti-particles, each of these 'as if' stories being constrained by the data we have, and each purporting to provide a plausible account of how those effects could have come about. But they could have come about in innumerable different ways from even the nicest story we can tell at any given time. And this seems to be the basic lesson to be learnt from the overthrow of immensely successful theoretical frameworks in the past, such as Aristotelianism, mechanism and Newtonianism. Of course, in retrospect we can see that *actually* none of these frameworks quite fitted the facts available even to investigators at their respective heydays. But then, as anyone who hears talk of yet another postulated fundamental force or anti-particle to set off the eccentricity of the latest postulated particle will realise, neither do our current theories. And even if a theory was produced that satisfied all the operational constraints (as Putnam puts it[12] and as Stephen Hawking apparently believes will happen), these constraints are only the constraints we are aware of in just the bit of the universe we have access to, and as we shall shortly see questions can be asked about that too. Moreover, for any theory that does fit all the data we have, there will in principle be alternatives that fit equally well. If anyone were to come up with such an alternative, then scientists would be in the position of Putnam's straight line physicists, the question as to which theory really told us how things are being undecidable.

But, in practice, we are very far from being able to construct even one adequate model for the data we actually have. There are currently all sorts of debates about the interpretation of quantum mechanics, let alone the truth or explanatory adequacy of such notions as the superposition of alternative states of affairs or the apparent ability of happenings in one system or part of a system to affect happenings elsewhere in the same system or even in another system altogether, or the claim that particles can have position but no momentum *or* momentum but no position. Mary Hesse does not exaggerate when she says:

> the fundamental theories of micro-physics are controversial and at present highly volatile. Nothing can rule out the possibility of a 'revolution' in physical theory that will reinterpret many identifications and laws of electrons, in a way similar to what happened to phlogiston.

But, as she goes on to say, there are two aspects to such revolutions:

> First they *leave alone* many local and particular truths, where we may as well say that we are referring to 'electrons', even though in a new theory electrons may not have other and perhaps deeper

> characteristics than we now postulate of them. *Hydrogen by any other name goes up in flames; electrons by any other name activate cathode ray tubes and geiger counters* (my italics). But secondly, the new theory may show that 'electrons do not exist', because our present theory about what they are and how they behave is found to be in detail and in general false. In just the same way 'phlogiston does not exist' because Priestley's theory is in detail and in general false.[13]

But, as Hesse concludes, false as an explanation of what since the acid experiments on the one hand, and the cathode ray tubes on the other, we have come to regard as *truths* of a factual, observational-cum-experimental sort.

So the best or best available explanation of any set of facts is not thereby true, and where, as in science, we frequently have no means of directly verifying what the explanation postulates, we would be very unwise to infer that our best explanations are true. Even the best scientific explanations available at any given time are always beset by difficulty in fact, and do not square absolutely with the facts, and even the best are always likely to be abandoned for some revolutionary alternative, postulating a quite different underlying picture of the world. (And even if these points were not true, and we had an entirely adequate account of the facts, it would not follow from its explanatory power alone that it was true.) Does this, then, mean that there is nothing in the claim implicit in any deployment of the inference to best explanation that the success of a theory in providing mathematically precise and pretty accurate predictions of the observed facts (including some as yet unknown) is no coincidence?

Empiricism

In answering this question, and in saying something about the 'some' extra 'truth' which Popper finds in cases where one scientific theory is judged to be superior to another, I shall draw on what Bas van Fraassen says about these matters in *The Scientific Image.*[14] Part of van Fraassen's considerable achievement in that book is that he has given an account of the scientific enterprise which shows how someone of a firmly empiricist frame of mind, who prefers to suspend judgment about the less observationally-determined claims of science, need not downplay the role of theorising in science. While rejecting the argument of inference to the best explanation, he is prepared to consider it seriously: it is, after all, the strongest weapon in the armoury of the scientific realist, and what underpins the pretensions of science to penetrate the inner essence of the world.

Van Fraassen begins his assault on the argument by pointing out that the demand for explanation of observed facts and regularities, if pushed back to any degree would result in the absurd conclusion that no scientific theory was ever satisfactory, because every explanation leaves us with some unexplained regularity (the unexplained behaviour of the current ultimate particles, or whatever). So there can be no general assumption that *explanation* is the supreme goal of science, as opposed, say, to measuring and recording data, predicting future regularities, and the like. Moreover an unlicensed demand for explanation, though undoubtedly implicit in the philosophical principle of sufficient reason, runs counter to the indeterminism of contemporary micro-physics, according to which the behaviour of individual particles is unpredictable. Hence, presumably, the philosophical predilection for 'hidden variable' explanations in micro-physics, a predilection which is not sanctioned by orthodox quantum theorists, however. More technically, but worth mentioning here, is that orthodox quantum theory also deals a blow to the idea that any observed correlations between distinct types of event must have some common and possibly unobserved cause. The Einstein-Podolsky-Rosen effect postulates that there can be correlations between independent events which have no common cause, and this effect – rather contrary to the intentions of its devisers – has now been corroborated in many actual quantum experiments. The most recent of such experiments was performed by Aspect who showed that while single photons emitted from a given atom will only randomly or unpredictably go through polarization filters to the left or right of the atom, if two are emitted simultaneously in different directions, and one happens to go through one filter (on the right, say,), the other will *randomly and without possible pre-programming* go through the other filter far more often than it should on the assumption that the behaviour of the two particles was mutually independent.

Van Fraassen does not wish to rule out all scientific searches for common causes of observed regularities or correlations. He fully admits that empirical discoveries can be made by postulating such things. But this is no argument in favour of using the inference to best explanation to the actual existence of a postulated but as yet unobserved common cause before it is observed. Reality is *as if* there are, for example, photons underlying both television pictures and the photoelectric effect used in doors opening automatically. So, the realist will say, there must be photons and doubtless the behaviour of photons in the two cases can be systematised mathematically in similar ways. Van Fraassen's response here is to say that so far all that the evidence entitles us to assert is that reality is *as if* there were photons obeying such and such laws. Only if explanation guarantees truth can we say more – and that is just what is at issue. But, is it not a miracle or cosmic coincidence if, as often happens, a theory not only explains phenomena, unifies so far unconnected phenomena and predicts new ones if it is not also true? Against the view that it would be an inexplicable coincidence if such a theory were not true (with all the underlying Aristotelianism about the basic adaptations of our minds to nature), van Fraassen proposes quite a different type of explanation, though also with biological overtones:

> the success of current scientific theory is no miracle. It is not even surprising to the scientific (Darwinist) mind. For any scientific theory is born into a life of fierce competition, a jungle red in tooth and claw. Only the successful theories survive – the ones which *in fact* latched on to actual regularities in nature.[15]

In other words, the success of our theories in predicting observable effects is due to the fact that science is highly competitive and only rewards such theories, so it is no coincidence that they get produced. And this picture, we may add, is more consonant with what we know of the history of science, of the abandonment of the ontologies and deep levels of past 'best' theories, and the non-convergence of successive theories at the level of deep explanation. It is also more consistent with our Darwinist picture of ourselves as somewhat adapted to our actual environment, but our minds no mirror of nature outside this environment, either in the very large or the very small.

Van Fraassen's own view of science is that science aims at (or should be seen as aiming at) giving us theories which are what he calls empirically adequate, that is which fit and anticipate observable natural regularities as well as possible. Theorising at deeper levels may well be an important part of the process of formulating such theories, and of discovering (observing) new regularities. The 'some' surplus of 'truth' which successful theories have over their competitors is just that they are more empirically adequate. Both history and logic should make us wary of claiming more, but, at the same time, we should not downplay or denigrate the value of greater empirical adequacy. After all, it is *this* knowledge which gives science its great pragmatic and technological power: the knowledge which, by virtue of precise reports and predictions of the actual empirical circumstances in which things will or won't happen, for better or worse allows us to intervene technologically in the world.

Ian Hacking has observed that the philosophy of science has been too long dominated by scientific theory, and by the grand theories of physics to boot, by what might be called the Einstein-Popper picture of science as *ascesis*, the solitary and detached search on the part of individuals of genius for nature's underlying secrets and structures, for its absolute essence, if you like.[16] But neither the history nor the philosophy of science can really support this quasi-religious conception. Mary Hesse has commented that the lesson of much recent philosophy of science is that what can actually be called empirical *knowledge* reduces to the pragmatic-empirical knowledge we gain through scientific endeavour of observed correlations and observational regularities, which form the basis of science's instrumental and pragmatic success, together with 'the hope that successful pragmatic knowledge is best acquired through the search for comprehensive theoretical systems rather than rule-of-thumb instrumentalism'.[17] That is to say, we have some grounds for hoping that we will do better at producing actual empirical knowledge by letting our fancy and imagination link and systematize what we know about the world as we see it by telling 'as if' stories about what the structures underlying the facts must be like or could be like, rather than simply, doggedly tabulating facts and observations without any guiding principles of organisation. And this is a reasonable enough hope, if only because fact collection and observation would otherwise be haphazard and directionless. But even if we produce some successful 'as if' stories we would be unwise to conclude anything about their truth beyond the empirical facts and regularities they allow us to unearth.

We should, in any case, realise what is often overlooked in abstract philosophical discussion of scientific theories, that is the ways in which the laws and theories of physics are frequently idealisations, approximations and simplifications. What is actually found in nature is far richer, fuller, more untidy, but we ignore 'irrelevant' aspects of the states of affairs we actually encounter and overlook mismatches of detail between what we observe and what our theories tell us to expect, putting them down to factors extraneous to what we are interested in, which we have been unable to control. We have already suggested a little of this in connection with Newton's theories, but Popper points out that Newton himself regarded the solar system as imperfect and likely to perish (and was therefore regarded by contemporaries as casting aspersions on the wisdom of the author of nature).[18] Popper goes on to quote C. S. Peirce, an experimentalist as well as a philosopher, to the effect that even the 'most refined comparisons' of masses and lengths, which far surpass in accuracy the precision of other physical measurements 'fall behind the accuracy of bank accounts', while the determination of physical constants 'are about on a par with an upholsterer's measurements of carpets and curtains'.[19] There seems, in other words, to be a tension between the untidiness of reality and the simplicity and generality we look for in our theoretical accounts of that reality. If these accounts were not reasonably simple and reasonably general, they would not be widely applicable, but the price of the desired simplicity and generality is an idealized account of nature, even at the observable level – an account which really applies only in theoretically ideal and simple situations, and which we bodge and modify in various ways, through bridge principles, *ad hoc* corrections and the like, to apply to specific actual environments.

In a recent collection of essays, engagingly entitled *How the Laws of Physics Lie*,[20] Nancy Cartwright gives plenty of examples of how the actual facts of nature are, as she puts it, fitted to equations, and a number of examples of how actual objects or situations do not obey the laws they are supposed to obey. Rather in the spirit of Peirce, she writes of Maxwell's mathematical treatment of the distribution of gas molecules in the radiometer, which is a little windmill in an evacuated bowl, whose vanes rotate. Maxwell's formulae are supposed to apply where in the radiometer there are inequalities of temperature and velocity, and where the viscosity varies 'as the first power of the absolute temperature' (Maxwell's words). But, claims Cartwright, these conditions are not met.

> in any of the radiometers we find in the toy department of

> Woolworth's. The radiometers on the shelves of Woolworth's do not have delicate well-tuned features. They cost $2.29. They have a host of causally relevant characteristics besides the two critical ones Maxwell mentions, and they differ in these characteristics from one to another.[21]

We know of no laws (or bridge principles) that would allow us to show how these other causally relevant characteristics determine the variations Woolworth's radiometers manifest from Maxwell's function for the distribution of gas molecules in radiometers. Moreover, not all the radiometers that do meet Maxwell's conditions obey his function; 'most have many other relevant features besides'. Cartwright concludes that in the absence of any laws linking the ideal case to real cases, Maxwell's distribution function has to be regarded as a pure fiction with no real explanatory force – a mere property of convenience, which we have no idea how to apply outside 'the controlled conditions of the laboratory, where real life mimics explanatory models'.

Cartwright's view is that what she calls models, which, roughly, are fictional theoretical descriptions of phenomena, are used in science whenever we want to apply mathematical theories to reality (something, remember, which is deeply embedded in the ideology of modern science). But these models never exactly fit reality. Their equations do fit the objects postulated in the model, but this is because the models are made to fit the equation. Different and incompatible models may be used for different purposes, and Cartwright cites a text on quantum optics which mentions the various models available for lasers: idealized, interacting models, soluble models, simplified dynamic models. In each case, the same phenomenon is explained in a different way and we make our choice of model depending on the properties of the phenomenon we are interested in. But

> in general, nature does not prepare situations to fit the kinds of mathematical theories we hanker for. We construct both the theories and objects to which they apply, then match them piecemeal onto real situations, deriving – sometimes with real precision – a bit of what happens, but generally not getting all the facts straight at once. The fundamental laws do not govern reality. What they govern has only the appearance of reality and the appearance is far tidier and more readily regimented than reality itself.[22]

This last quotation from Cartwright could perhaps also be applied to the work of the various philosophers of science I have been considering in this chapter. At any rate, I do not want to pretend that they will all agree with each other, which is clearly not the case. Popper and Watkins for example, would disagree with the 'anti-realist' tendencies of all the others I have mentioned, their tendency, that is to downplay in various ways the pretensions of science to give us literally true descriptions of reality. Kuhn and Hesse, as we have seen, disagree on the independence of observation from theory. Cartwright is far more sceptical than van Fraassen appears to be on the empirical adequacy of theories. On the other hand, Cartwright and Hacking are more confident about the existence of theoretical entities than van Fraassen, where those entities can be seen to play a genuinely causal role in experiments. Kuhn appears at times to cast doubt on the rationality of scientific procedures, and appears ambivalent about the growth of scientific knowledge, something all the others see as a basic datum to be explained. And there are naturally many other disagreements and differences, which are not so central to our concerns here.

Despite all the differences just mentioned, though, I do think that there are – despite appearances to the contrary – some fundamental and relevant points of convergence. Thus, for example, we find Popper himself frequently warning us against thinking that our deep explanations are true or even probable just because they are confirmed at the observational level, and it is Popper more than anyone who emphasized the revolutionary change of theoretical framework in the transition from Newtonian to Einsteinian physics. Popper also wrote in his early *Logic of Scientific Discovery* that 'the various ideas and hypotheses (of science) might be visualized as particles suspended in a fluid. Testable science is the precipitation of these particles at the bottom of the vessel.'[23] According to Popper a theory is scientific only when testable by observation or experiment, and scientific choices can be made only between scientific theories. But, as we have already seen, from the point of view of van Fraassen this can all be analysed in terms of greater empirical adequacy, and as Popper in effect admits, despite his occasional flirtations with the inference to best explanation argument, this tells us nothing about the truth of the 'higher metaphysical reaches' of empirically adequate theory.

Kuhn, in contrast to Popper, is often regarded as a sceptic about science, but such scepticism as he has is largely (and correctly) directed against the grand claims of science to provide us with true pictures of the world beyond anything that could be empirically tested. And this scepticism is shared by Hesse, Hacking, Cartwright and van Fraassen, despite their differences about the limits of observation and the extent of empirically based knowledge. For all of them, as arguably for Popper in the end, what really and positively grows in scientific knowledge is *positive* knowledge of the world: that which is revealed in our observations and experiments – and which in various guises survives theoretical and conceptual shifts – rather than through theoretical speculation about the observationally unattainable essence of the world.

The absolute conception of the world

It is time to contrast the conception of scientific truth we have now arrived at with the view of science as aiming at what might be called an absolute conception of the world, not tied down to our or any other specific manner of perceiving it. The absolute conception of the world is taken to be a conception not essentially dependent on the viewpoint or sensory equipment of particular observers. It aims at revealing the inner structure of the world as it is in itself, and in a way which would be intelligible to any intelligence, whatever its viewpoint or sensory equipment. In revealing the essential nature of things, it would also explain how it is that particular types of observers came to have the viewpoints they have, and how their particular forms of perception were related to the underlying structure of reality, the underlying elements of reality being taken to be those that were really responsible for the causal processes and regularities in the world. In a sense the absolute conception of the world, for one who possessed such a thing, would be like contemplating the world from outside – a god-like or decentred view, perhaps, such as was envisaged by Laplace when he spoke of an intelligence able at a given instant of time to comprehend everything in the universe from the largest star to the smallest atom 'in one single formula' and to see both past and future as certain.[24] Laplace speaks of our knowledge of the solar system as affording some feeble insight into the nature of such an intelligence. It is clear enough, though, that the intelligence involved would, at least for the purpose in question, be conceiving itself as outside the system it contemplated in this way. If it were regarding itself as inside the system, it would have to regard its own future mental activity as part of the system and part of what future states of the universe would be based on. But it could hardly be thinking now what it would think in the future, or its future thoughts would be its present thoughts.[25]

It is easy to see why a conception of the world from as it were outside, should have been taken by Wittgenstein in 1916 to have strong ethical and even religious overtones. Now while there is nothing theoretically wrong with entertaining the possibility of such a conception, to take it as an aim of empirical science and to see empirical science as affording the possibility of such a conception for creatures located within the system requires that we base our science on the assumption that the perceptions of any particular observers within the universe can be fully explained by recourse to the primary qualities of things. This requirement is necessary because otherwise, if we have to bring secondary qualities into the explanation of our perceptions, the account of the structure of the world which we give will depend crucially on how the world appears to us, and such an account will not be fully graspable by intelligent beings lacking our sensory equipment. At the same time, an absolute view of things dependent on no particular sensory viewpoint on the world will give a clear distinction between those properties which are in things and those which are only due to the interactions between things and particular sensory faculties – and this Newtonian and post-Newtonian science aims to do.

In the last chapter, I argued that, despite attempts to do so, the attempt to explain our perception of the world in fully objective terms fails. This aspect of the absolutist programme (which is, of course, still believed by many physiologists and psychologists) breaks down at the crucial point; perception itself can't be explained, and lacking such an explanation we have to remain sceptical about the further claim that the world really or fundamentally contains only those properties mentioned in the theories of science. Certainly, an explanation of the world (in any terms) which did not explain sensation and perception would be radically incomplete because such mentality is undoubtedly part of the world. Bernard Williams, for example, makes it a condition of any purported absolute conception of the world, that it does, at least to the sensory beings involved, non-vacuously explain how they came to have the perspectives and

sensations they do have in terms of 'non perspectival materials available to any adequate investigator, of whatever constitution'.[26] Williams believes that evolutionary biology and the neurological sciences have provided some non-vacuous explanation of human perspectives in absolute terms, and while this may be so to some degree, they have not done so when it comes to the crucial question of conscious perception itself. This being so, any purported absolute view of the world currently available to us fails to explain how we ourselves come to be in a position to frame it or any other conception of the world, and this would seem from our point of view at least to destroy its title to completeness, and hence to absoluteness. From the other side, of course, the very theories we produce – ostensibly delivering an absolute view of the world – are themselves a product of the very mentality and conscious experience they cannot explain.

The passing of time

The absolute conception of the world seeks to give an account of the world which would in principle be equally accessible to creatures at any point in space and time (or outside space and time altogether). The Special Theory of Relativity, however, shows that because of certain physical facts about the universe, any absolute conception of the world will end up by making temporal orderings of events belong to what Bertrand Russell referred to in this context as 'the subjective part of our observation of physical phenomena', rather than to 'the objective part which is to enter into physical laws'.[27]

If events are to be ordered absolutely in time, as being absolutely speaking prior to, simultaneous with, or later than all other events, we will need to postulate some absolute frame of temporal reference against which events can be timed or dated. This might be provided by postulating an absolute time and space against which bodies could be said to be moving and changing absolutely rather than merely relatively (which was Newton's view of the matter), or something static and unchanging within space and time, against which movements of bodies and temporal positions of events could be measured. There are obvious difficulties involved in the detection of an absolute time and space, and in the nineteenth century the most popular solution to the problem of temporal measurement was to postulate the existence of the ether, a stationary stuff supposedly permeating the whole universe, against which time and change could in principle be measured.

But the Special Theory of Relativity postulates first that no ether can be detected, from which it follows that all motions and velocities must be regarded as relative to the motion and velocity of the measurer, and second that the velocity of light is always constant relative to an observer, whatever the relative velocities and directions of the light source and the observer.

From the two postulates of the Special Theory a number of consequences follow, the most important of which for our purposes is that we can no longer speak of events being simultaneous in any absolute sense.

The point here is that because the light rays from any event take a certain time to reach an observer of that event, observers situated differently relative to that event will see that event at different time intervals from its occurrence. If we now consider two widely separated events, A and B, and three widely separated observers, 0_1, 0_2 and 0_3 at different distances from each other and from the two events, the possibility now arises that 0_1 will see A and B happening simultaneously, 0_2 will see A happening before B, and 0_3 will see B happening before A. Given that there is no absolute frame of reference in which A and B can be ordered, we have to conclude that there is no reason for preferring 0_1's version or 0_2's or 0_3's. 'The order of events is a function of the observer's position and velocity relative to all other observers. Simultaneity is a relative matter.'[28] In other words, any experience of the universe given to creatures in the universe, because of the physical conditions under which information about events can be gathered, will order events according to the position in the universe of the creature doing the ordering, and there is no temporal ordering of events that is, absolutely speaking, to be preferred to all other orderings.

The Special Theory of Relativity does not imply that our experience of time is a pure illusion. Indeed, even in cases where, owing to the speeds and distances involved, two observers would actually reverse the temporal orderings of two events, how they would order them would be perfectly predictable given the facts about the relative velocities, the relevant distances, and so on. Our actual temporal orderings are, in other words, absolute enough once they are relativised to the particular reference frame in which we are making them, and indeed can if necessary be corrected by reference to that reference frame. Relativity theorists, moreover speak of a function calculated on the basis of the distance and time between two events, which is called the 'interval' between events, as being within certain contexts invariant between observers and a genuine physical relation between the events.

A question, nevertheless, remains as to the livability of a conception of the world in which the experience of the passing of time is seen as inherently perspectival in this way. We seem to be avoiding the problems involved in the subjectivity of our temporal orderings only by, as it were, standing outside all temporal perspectives and seeing each as a function of the perspective of a particular observer. We are, so to speak, seeing all events from a timeless point of view, as, in a sense, all simultaneous. Events are objectively located in a relativistic space-time, whose division into separate spatial and temporal dimensions will be seen as depending on one's point of view.[29] Bodies are not seen as moving and changing through time, so much as worms embedded in a motionless, unchanging space-time, like fossils in a rock.

And this is not a possible perspective for us to live by, given that our lives and experiences are fundamentally temporal. The transition from past to present and future is one of our basic forms of experience. Even if we like to adopt a timeless view of the universe, in which temporal succession as we experience it is seen to be simply a function of our perspective on the universe, analogous to secondary qualities in the Lockean account, our own experience and apprehension of the timeless view would itself be in a temporal relationship to that view, as our experience developed in time. In our experience, there is a succession of experiences, in which we remember past experiences and speculate about future ones. This is a *real* succession, for the successive experiences can't be combined. The past ones cease to exist and the future ones do not yet exist, and they are not all equally present in some space-time manifold. So even if, in some absolute sense, our temporal experiences of physical phenomena are not part of the objective part of those phenomena, there is within them a *real* succession of moments, with some ceasing to be and others not yet in existence.[30] The relativistic a-temporal perspective of scientific theory is one which is belied by our very act of considering that perspective. While on the one hand we are saying that our perception of temporal relationships is a function of mere perspective within relativistic space-time, that perspective itself conditions the context within which we think and deliberate about such things.

Some would draw a different but, if correct, equally unsettling conclusion for the absolute view of the world from certain aspects of quantum theory. The claim is that the physical process of observation so affects the reality being observed that we cannot talk of an objective reality at all. According to Heisenberg the assumption that we can divide the world into subject and object 'is not permissible in atomic physics; the interaction between observer and object causes uncontrollable large changes' in the system being observed, while Bohr claimed for similar reasons that 'the finite interaction between object and measuring agencies ... entails the necessity of a final renunciation of the classical ideal' of objectivity.[31] These ideas of the Copenhagen school have become very much part of received wisdom in popular science, but I do not want to rest anything in my critique of the absolute conception of reality on them. For one thing, the so-called Copenhagen interpretation is not the only interpretation of quantum theory, which, as already remarked, is widely regarded as a highly controversial area. More to our point, the *argument* of Heisenberg and Bohr as stated shows at most that we interfere with reality by observing it and that this interference makes it impossible to discover what the precise state of reality was before being observed: it does nothing to show that there is no reality we interfere with. That we interfere with reality by observing it and that this interference is significant when observing the very small is something one can readily accept, given that observation is a physical process, but even this would not by itself show that all observers might not be able to come to a general agreement about the underlying nature of the reality that their various observations and probings are probings into, and even to some imprecise knowledge of its state prior to being probed. The Einsteinian denial of absolute simultaneity, and the consequent adoption of a basically tenseless view of the world, seems firmer ground for questioning the attainability by creatures within the world of an absolute conception of the world which is also a complete account of their experience of the world and something they could live by.

Science and mythology

Aside from problems involved in fitting conscious experience into the absolute conception of the world into the scientific treatment of time, we have arrived in this chapter at a somewhat less exalted view of the claims of scientific theory to provide us with deep truths about the physical world than that with which we started out. The truths science delivers and in terms of which science grows are largely truths about the world which observation and experiment make available to us. In that sense, they remain truths with a firm basis in our viewpoint on the world, and this is what is valid in the empiricist view of knowledge as requiring a basis in experience. Apart from delivering empirical and pragmatic truths, which may certainly involve corrections of our original experientially derived conceptions of the world in the name of a greater objectivity of view and which will be indifferent to the way things affect us humanly speaking, science also presents us with pictures of what the universe as a whole is like.

It is with these claims to present us with accounts of the universe as a whole, which apply everywhere and may even presume to take up a stance outside the universe and which also probe deep beneath the surfaces of things, that science goes beyond anything which might properly be called knowledge and enters the realms of metaphysics and mythology. It is not required for our finding wide-ranging and extensive regularities in our corner of space and time that these regularities apply everywhere, nor does the successful discovery of such regularities imply that they apply everywhere. It may well be part of the ideology of science that the universe is governed by simple, discoverable laws but this cannot be taken for granted. Indeed, we have found reason to doubt that things in everyday experience conform exactly to the scientific models we propose. The world may be less regular than a universal science supposes and the objects in it more fuzzy than its laws envisage. As far as anything we are justified in claiming goes, to quote Mary Hesse again,

> all we can say is that science depends on believing that there are objective local regularities among physical entities and processes and that we can learn some facts about these regularities independently of our wishes and conventions. The accumulation of such facts constitutes scientific progress.[32]

And, I would add, constitutes scientific objectivity, in contrast to the world of human art and culture, which reflects human wishes, albeit in a sublimated form, and depends on human convention.

And what of the rest of science, the grand, even inspiring attempt of scientific theory to answer deep questions about the nature of the universe, of matter, of time, of reality itself? Again, Hesse seems to me to have it right:

> With regard to such questions, scientific theory is indeed constrained in detail by the requirement of 'empirical fit', but the content of what it says has otherwise no truth-bearing privilege over other kinds of metaphysics or mythology. It does indeed itself have many of the characteristics of mythology in our society.[33]

The empirical fit of science is, as we have seen, tied down to our viewpoint on the world, as extended to be sure by our instruments and technologies, but our viewpoint nonetheless. The mythological role of science in our society is, moreover, liable to be destructive of human sensibility in that methods of science quite properly abstract from human wishes and conventions. Our survey of the limitations of science in establishing its own credentials to provide us with deep truths about the world or even to explain our own experience show conclusively enough that describing science as mythology is not mere rhetoric.

Lest this conclusion seem a more negative appraisal of scientific endeavour than I intend, let me emphasize once more that the scientific aim of an absolute, impersonal view of the world and the undoubted success of science in extending our knowledge of the world is a testament to the range and power of human imagination. For there is a world beyond us and beyond the specific explicit disclosures we make of it, and it is morally admirable to submit ourselves to a reality outside ourselves, as scientific research requires. In attempting to probe the beyond we are always, as Marjorie Grene puts it, 'in advance of our data in the aspiration to knowledge' yet at the same time 'caught in the ambiguity of our finitude'. There is indeed

> no absolute, once for all, knowing by human beings ... There are only ourselves, using all the means at our disposal: bodily

> orientation, sensory images, verbal formulations with all their over- and undertones, social taboos and imperatives – including all the lore of practice and procedure of any given discipline we have been trained to – and, finally, our deepest, widest vision of the world we dwell in.[34]

This is beautifully expressed, but I would qualify the last phrase by pointing out that the deepest, widest vision remains human, though not in itself all too human: the all too human comes in with the hubris of believing that we can transcend the human through scientific theorising and technological manipulation of ourselves and nature. Hubris is also implicit in the belief that science and technology are the only truly serious elements of our culture, and that they alone can give us the 'deepest, widest' vision of ourselves. Despite all their qualities, science and technology cannot be the central elements in a healthy human culture. The arts, by contrast, offer us ways of understanding ourselves and coming to terms with our potentialities which take us far nearer our inner essence than science and technology can.

CHAPTER 4

Human Culture and the Role of Artistic Expression

The conception of the world projected by modern science is one that denies our naturally anthropocentric attitude to existence. Even if, as we have seen, natural science is not ultimately successful in cleansing its observations and theories of anthropocentricity, and it remains we, as human beings, who in science observe the world from our point of view, and impose our metaphors and metaphysical constructions on the phenomena, there is still a clear sense in which natural science does allow us to transcend our initial limitations of perspective in order to frame a broader conception of the world. In particular, science presents accounts of our perceptions and of our behaviour as products of a more inclusive set-up, in which what we perceive and do is presented as the causal effect of circumstances prior to and outside us.

How far such accounts are or could be successful in explaining either our actions or our consciousness is a matter of dispute. My own feeling is that no account of consciousness of which we have any grasp could even begin to explain how electro-chemical impulses in the brain might not only stimulate activity in an organism, but also become somehow transparent to that organism, as felt perceptions of blue, or red, or pain. That this happens is indubitable, but explaining it or even knowing what form an explanation of such a transformation of quantity to quality should take is an altogether different matter. Furthermore, no detailed account of human action has been offered, which would allow us to predict fully my behaviour over the next hour or two. Human behaviour can actually be predicted only in very controlled circumstances, or on a very broad canvas. So there is plenty of empirical slack between what natural science can actually do regarding human consciousness and behaviour and what it quite rightly sets out to do in proposing a scientific account of human activity. Nevertheless, it is clear what such an account would be attempting to do: namely, to produce the same sort of displacement of the human agent from the centre of his world as, say, astronomy has done to the geocentric world-view of Aristotle and Ptolemy.

Such a displacement would, if it were successful, have a more profound effect on our picture of ourselves than did the transition from the geocentric world view, for it would be finally destructive of the concepts of the human self as initiator of its activity, and of freedom as central to our evaluations of human activity and basic to the relationships we conceive ourselves as having, concepts which, though metaphysically obscure, are surely central to human life and activity. The displacement of the values of our human world by an advance of natural science into the human world would indeed provide the final vindication of Nietzsche's claim that it is *error* (about the self, about freedom) which has transformed animals into men.[1] Nietzsche goes on to ask whether truth is capable of transforming man again into an animal. Underlying the characteristically overblown rhetoric is the central problem raised by any philosophical examination of the relationship to mankind and the rest of nature. We can travel some considerable distance in rebutting the Nietzschean-scientific challenge by considering the forms of life made possible by the 'error', and, seeing how these forms of life distance us from the animal within.

Human culture and practices

Human beings are part of the world of nature and share with other animals biological drives and needs. Birth, death and copulation frame our lives; within the frame we need food, warmth, health and companionship, as do other animals. The fact that we all, as humans, share a common frame in this way, rooted in our biological nature, is what makes cross-cultural communication possible. What, though, is distinctive about human beings (and perhaps about some other animals, too) is that we are not conditioned merely by these biological needs. Our lives are not simply strategies for survival or devices serving the interests of our genes. Such are no doubt the effects of (some of) the actions of some of us; if it were not so, then neither individuals nor the species as a whole would survive. But this truism cannot be taken to characterise most of what we do, at least when we look at what we do from the point of view of the agent himself, which is the only point of view from which we can organize and experience our lives. Being self-conscious, we are capable of evaluating our experience and activity for its own sake. We are capable of modifying our aims and goals in the light of such evaluations, of learning from and correcting our mistakes. I would see our possession of consciousness, as discussed in chapter 2, as central here. For we do not simply revise our activity in the light of goals we are given. We are conscious of having certain goals, of those goals as being not just given to us but our own goals, part of what we are for the time. And we can change and alter what we are in the light of our self-awareness.

We are capable of finding meaningfulness in our lives and activity which goes beyond the straightforward satisfaction of purely physiological need and biological impulse; and this may indeed be based in the way we have a conception of ourselves as existing through time, and so capable of being haunted by the past (to use a phrase of Mary Midgley), and of seeking continuities with the present and the past in the future. Take, for example, the way a meal, something very directly concerned with instinctual needs, may not simply be a means of assuaging hunger, but can become an elaborate and enjoyable focus of conversation, in which the taste of the food and the bouquet of the wine are treated partly as ends in themselves and partly as the setting for – shall we say? – edifying discourses. (It is, I think, significant that the highest rhapsody to philosophy, Alcibiades' speech in Plato's *Symposium*, is represented as taking place at a meal.) Or take the way a road into a city may be either a dead space in which journeys are simply dead time, a means of getting from place to place, or, as town planners from the Renaissance Popes to the nineteenth century fully understood, something that can be constructed with artefacts and vistas to delight and impress, so that what would otherwise be dead time becomes full of shared meaning and value for the traveller. Or, as Hayek has pointed out, consider the way in which farmers have traditionally been guided by rules whose outcome guarantees less the ability to consume and produce much, rather than recognition by their fellow-practitioners as good farmers, with a fine reputation.[2]

These examples illustrate the ways in which a man's journey through life is transformed from being nothing more than an instinctually dominated transition from the emptiness of non-existence to the emptiness of death. This transformation is effected through the human power to evaluate experiences and adopt goals in ways that have nothing directly to do with the dictates of instinct or raw impulse. The exercise of this power by an individual crucially depends for its enactment on the prior existence of what Alasdair MacIntyre[3] has referred to as practices: activities (such as farming, politics, game-playing, academic study, poetry, music, the law, marriage, the army) which are structured in such a way that an individual's entry into a practice and his finding of meaning in it is conditional upon his accepting as binding on him the norms, ends and disciplines presupposed in those practices. The drunkard and the glutton lose what food and wine have to offer because they allow desire to destroy the delights made possible only through discrimination. Nero had himself crowned with the poetry prize at Olympia in AD 67, but that did not give him the satisfaction of being a good poet – for that he would have had to have submitted his work to the unforced judgment of those who really understood fineness in poetry. The point I am urging here is that the human ability to escape from the tyranny of instinct and physiological need and greed depends crucially on the existence of cultural forms with pre-existing standards, which propose to us goals which are not co-extensive with individual whim or instinct. Even if an individual has the power to crush the wills of others, as Nero undoubtedly had, in crushing the wills of the Olympian judges he deprived himself of the satisfaction of being truly recognised as a fine poet. This is a satisfaction which depends essentially on a cultural form, and is made possible by that mutual recognition which comes from an acceptance of values and standards regarded as binding on both parties. This is something ruled out by the nature of the case in a pure master-slave relationship, as Hegel recognized. The slave bends to the power of the master, but does not respect him. Honour, which the master seeks, as a confirmation of his worth and self respect, presupposes that the slave recognize him as honourable; and this in turn requires that both master and slave recognize certain qualities, characters and actions as objectively honourable.

Michael Oakeshott has written: 'Human beings are what they understand themselves to be; they are composed entirely of beliefs about themselves and about the world they inhabit.'[4] He goes on to say that we inhabit a world of 'intelligibles', of occurrences which have meaning for us, in which our wants are not perceived as biological impulses or genetic urges, but as 'imagined satisfactions' which have reasons rather than causes, and which can be approved or disapproved, chosen or rejected. Paraphrasing Marx, he says that man is 'what he becomes; he has a history but no "nature".'

Oakeshott's conception of human life is one of self-conscious agents choosing their actions against a background of culturally mediated forms and practices. Our self-understanding is in terms of how we locate ourselves within those forms and practices. This is an active and creative process demanding from each individual some sense of what the practices of his culture demand, and some response from him to those demands. (The type of activity and creativity envisaged obviously implies the free, metaphysical self natural science would obliterate.) Not to have a sense of the demands of those practices is to be 'not a human being, but a stranger to the human condition'.[5]

Oakeshott's conception of the human being as one who is composed of the beliefs he has about himself may appear at first sight to be similar to the Sartrean view that human beings have no essence and find themselves only in repudiating their past commitments, or even to the utilitarian image of man as simply a bundle of satisfactions and dissatisfactions, manipulable as whim, desire or science dictates in order to increase the total sum of satisfaction for each individual. On these views, everything is permitted, providing (in the case of Sartre) I freely chose it, or (in the utilitarian case) it increases the number of momentary satisfactions. Past commitments have no intrinsic value and are always, as it were, up for renegotiation. One standard answer to these positions is to invoke man's biological nature or essence,[6] as an embodied being, with certain needs and satisfactions desirable by considering his biological nature, but this is a response that is not open to Oakeshott, nor do I think that we can simply rest our conceptions of human life on our biological nature.

While I wish to reject Oakeshott's crude separation of the natural and the historical in human life, he is right to emphasize the extent to which in the human world, the satisfactions and meanings available to us are not purely biological but are culturally mediated and expressed, and presuppose on our part aims, intentions and conceptions of who we are and what we are doing. It is true, and important, that there are, in any human life, certain universal features, basic needs, abilities and propensities. These include, as already noted, need for food, warmth and companionship, propensities to bonding and altruism and aggression and to other emotions of various types, to laughter and dance and music. Many of these may be based in biological facts about us, and this universality of action and need is what makes cross-cultural communication and evaluation possible. Oakeshott appears to underestimate the significance of this common biological basis. But biology does not determine the way basic needs and propensities develop, nor the forms they take, which have developed in manifold and diverse ways in different human societies. As Schiller points out,[7] there is a great chasm between 'unco-ordinated leaps of joy' and the order of a dance; and there is all the difference in the world between an animal or bird 'singing' and responding to signals, and someone enjoying that singing for its own sake. To that extent, human beings are, in Oakeshott's terms, 'historical'.

While certainly part of nature, our natural needs and desires and propensities are always expressed in and through cultural, historically developing forms. But it is because of the historical roots of the cultural and institutional forms against which we understand who we are that the Sartrean *acte gratuit* or the utilitarian manipulation of desire are to be rejected. To quote Oakeshott again

> being human is recognizing oneself to be related to others ... in virtue of participation in multiple understood relationships and in the enjoyment of understood, historic languages of feelings, sentiments, imaginings, fancies, desires, recognitions, moral and religious beliefs, intellectual and practical enterprises, customs, conventions, procedures and practices; canons, maxims and principles of conduct, rules which denote obligations and offices which specify duties.[8]

In other words, the Sartrean can only make an ultimately vain and vainglorious attempt to efface whatever identity he has by means of an act intended to sever himself from the links which constitute what he is, while utilitarianism in its assumption of the endless manipulability and malleability of desire overlooks the way in which our very desires, let alone our sense of who we are, are constituted by our being rooted in a particular culture, with particular relationships, in a specific time and place.

Common culture and religion

Following Oakeshott and MacIntyre, I have been considering human culture in terms of the practices and relationships which a man is born into or into which he enters during his life. I am suggesting that it is in terms of the relationships and practices into which he is born that a man first comes to some notion of who he is and where he has come from. As he moves through his life, he will embark on new relationships and engage in new practices, constructing for himself some sense of who he is becoming and where he is going. Some, though, will want to say at this point that human culture – or at least a genuinely common culture – is not just a collection of relationships and practices, such as the family, the law, the army, poetry, cooking, sports, farming, and so on, but a whole integrated framework into which all these disparate activities fit, and within which individual lives take on a global significance, as narrative, quest or journey. The stages of life of a Hindu, or an ancient Spartan, or the Christian's journey through this vale of tears form examples of what I mean here, the point about a *common* culture being that public rites and acknowledgements of the various stages of the journey and the emotions appropriate to the various incidents of life take an individual out of his own narrow concerns and infuse his life with the meaning which an objectively determined public realm has bestowed on it, and which typically forges links of meaning between the living and their ancestors and descendants.

Obviously the prime example of the sort of overarching practice which I have in mind here is religion. For believers, religions provide precisely those links between the living and the dead and the not yet born which enable men on earth to see their activities here as more than meaningless moments in the history of a transient, chance-ridden biological species, to be valued only for whatever passing pleasures or satisfaction they bring us. The crucial sense that a religion gives to its adherents is the sense that nothing is lost, and that the passing of time and death itself can be overcome. The redemption of what has been lost in the past, and the connection that we have with what will happen in the future is symbolised in Catholicism through the ritual of the Mass. All Masses are the one Mass, originally enacted by Christ at the Last Supper and on the Cross, but looking forward to and encompassing all the future performances of the rite. In the Mass, time is overcome; here time becomes space, as Wagner's Gurnemanz puts it in leading Parsifal into the Chapel of the Grail, in that past time and future time are brought together in the timeless, because eternally renewed and eternally renewable, Grail ritual. And if our self-consciousness gives us possibilities of evaluating our activities and setting goals for ourselves which enable us to follow and construct cultural forms, to be guided by reasons in our activity rather than impelled by biological causes, the very standing outside the temporal flow which makes such self-conscious planning possible also brings us the realisation that what is past is gone, and that each one of us will eventually die. Whether we think it is irredeemably gone and that death is the end for each of us of all our efforts and projects – and all that that implies about the meaning of our lives – will, of course, depend on our religious convictions. What is clear, though, is that the religious urge is no mere by-product of transient economic conditions, but is intimately linked to our nature as self-conscious and reflective, but at the same time material, embodied and mortal.

Whatever individuals in contemporary western society might individually believe about religion, it is obvious that religion has lost the central unifying role it once had. We no longer inhabit an order in which the Christian framework can be taken for granted in public discourse. The full mutual recognition and corroboration of feeling which a universally shared framework of religious belief can provide to individuals through its ceremonies, symbols and myths is not available today even to the religious, who are in a minority in our society, and who have it brought home to them all too often that the majority are at best indifferent and careless towards what they hold dear. (Their predicament is made even worse by the fact that many of their own clergy manifest a similar disregard for fundamental points of doctrine.)

I do not want to say that all here is loss. There are negative aspects to religion, particularly when it is strong, confident and dogmatic. These include a somewhat restrictive attitude to human thought and inquiry. Some, too, might feel, like Horace (*Odes*, Bk III, no. 29) that the truly joyful man is he who can say 'Today I have lived' – because no God or future event of any sort can remove that from us; and that there is something claustrophobic about an existence constrained by divine prescriptions. On the other hand, although Nietzsche would certainly have found a divinely circumscribed existence claustrophobic, in *The Birth of Tragedy* he regarded the cheerfulness of

Horace's Ode that of the slave 'who has nothing of consequence to be responsible for, nothing great to strive for, and who does not value anything in the past or future higher than the present'.[9] And it is true that so far as notions like honour and integrity play a meaningful part in human life, concentration on momentary pleasure is bound to be perceived as grovelling and dishonourable.

Whatever one might say about Horace's *carpe diem*, for better or worse, we today are more reflective, more self-conscious about what we are, about how we are formed by and through our culture and history, and what we might become, than were people in, say, medieval Europe. What I now want to suggest is that one vital way of understanding ourselves, our nature and our history is through art and the study of art, and further that the particular type of self-understanding afforded by the arts is one which was still necessary even within a religious context. But for us today, without a shared background of belief to form the foundation of our self-understanding, the need for some sort of self-understanding from within – which I hope to show the arts can provide – will seem particularly urgent just because we have no transcendently guaranteed framework telling us who we are. (I shall, in chapter 6, consider how far a vibrant artistic tradition depends for its survival on the shared symbolic order religion can provide.)

A la Recherche du Temps Perdu

Inspired by the example of Proust, one might feel strongly the temptation to speak at this point of art itself as redemptive, as capable of doing in its own way what religion does in redeeming us from death and the passage of time. Certainly in works of art the conception of the artist is concretized in enduring form; more specifically, in literature and music and, in its own way, in the visual arts, a significant unity is created between the parts of the work. In music and literature, there is a clear sense in which we are presented with a narrative, paralleling the religious idea of life itself as a narrative. Proust, though, meant something far more specific than any of this. For him, the tragedy of life was the loss of the past. The redemptive power of unconscious memory (unlocked initially in the tasting of the madeleine) was precisely that it showed that the past was not lost, but present in memory if only memory could be unlocked. This actually led Proust to a rather bizarre evaluation of works of art, according to the power they gave to each individual to revive his past. So Vermeer's little patch of yellow was for the narrator in *A la Recherche* on a par with the 'little phrase' of Vinteuil (originally a somewhat anonymous phrase from César Franck, apparently), or even, presumably the madeleine itself. Proust led a fully social life until the age of 38 (soon after his mother had died); the remaining thirteen years of his life following his own mystical experience with the madeleine were spent largely in seclusion in his task of redeeming the past in his great novel, and he died, apparently intentionally, on completing his final volume at the age of 51.[10]

Saintly though Proust was in his devotion to his task, his theory of the redemptive power of art suffers from a number of serious defects, apart from the bizarre attitude to art it led him officially to espouse. The redemption Proust envisages is one that comes from simply reproducing the traces the past has left in his mind. In effect, he is denying the role of the imagination in creating a work of art; officially for Proust the ideal artist can be no more than a transcriber of memory. Against this, it should be clear that any piece of descriptive writing or painting requires selection, abstraction and emphasis on the part of the writer or artist. Proust's own particular genius allowed him to focus on particular aspects of feeling, experience, and character, and to express them with such concrete vividness that things we all know and have experienced only inchoately are brought before us in total clarity, fully articulated. It is easy to see how *he* should have seen his method as simply the reviving of lost memory, but in doing so he overlooks the imaginative power that so characterises his work. Furthermore, it would be hard to regard more than a fraction of the art and literature that we have as primarily an evocation of memory; a transformation of experience by means of imagination, maybe, but not an attempt to redeem one's own past by reconstructing it in concrete form.

The 'two' cultures

So the temptation to think of art as directly redemptive in Proust's sense must be resisted. Art cannot replace religion in that sense. What it can and frequently does do, though, is to elaborate our understanding of ourselves and our history, and it does so in a way that science cannot and even, as I shall suggest, religion itself cannot. The contrast between the scientific and artistic approaches to experience was the real substance of the much misunderstood debate between Snow and Leavis on the 'two' cultures. One of Leavis's main points was that Snow betrayed insensitivity to the nature of the human world when he spoke of science and technology as constituting a culture. As we have already seen, science aims at producing conceptions of the world which prescind from human sensibility and concerns. It also aims at being value-free in leaving aside all questions of the importance to us of its discoveries and concepts. It aims to construct classifications and explanations of natural phenomena which do not relate in any direct way to the way we perceive them or the way they affect us or impinge on us. There is nothing improper in these ideals. Indeed, the discipline implied in science of submitting oneself to the way things actually are is itself morally honourable, something surprisingly overlooked by Leavis. Moreover, the results of scientific study may have a deep effect on the way we conceive ourselves and our world, making, perhaps the world view of the *Divine Comedy*, say, something we can hardly even reconstruct today, let alone consciously adopt. But science itself is indifferent to the way its discoveries affect us, the way they impinge on the human world. If it were not, it would be in danger of losing the very title to objectivity which is the basis of its standing. While one's view of how the world is is part of one's culture in this regard, which is the pursuit of objectivity implicit in Western science an admirable part of our culture, science cannot itself constitute a culture, for a culture is precisely the elaboration of the way groups of human beings respond to the world and find meaning in it. By contrast, no one can actually live out the scientific perspective, in which nothing matters more than anything else, and I or we are not the centres of our world.

There is, of course, an attitude some people derive from what they take to be the world-view of science – a bluff and dismissive attitude to other people's sense of what is significant especially when it stands in the way of material and technological 'progress', and this, too, was part of Leavis's complaint against Snow. Encouraged by the technological success of science, there is, as Leavis suggested, a new barbarism abroad

> the civilized barbarity, complacent, self-indulgent and ignorant,

> that can see nothing to be quarrelled with in believing, or wanting to believe, that a computer can write a poem. In the interests of self-congratulation it simplifies social problems by eliminating the life: the complexities it reduces them to are mechanical, or treatable mechanically – it hates the organic, and its simplifications kill.[11]

But this is a case of the misapplication of scientific methods to areas where that sort of objectivity is inappropriate, which attempts to quantify what cannot be quantified, and dismisses as reactionary obscurantism all values and perceptions not in accord with the planners' conception of rationality, which fails, for example, to conceive that for some people organic squalor is preferable to utopian architecture.

In art, on the other hand, we are all the time concerned with the humanly significant. As Goethe put it in one of his posthumous aphorisms on art, 'we know of no world, but in relation to man: we desire no art but the art which is the imprint of that relation'. For Leavis, too, literature and the study of literature was central here, not in order to establish purely literary values (whatever they might be), but precisely because 'the judgments the literary critic is concerned with are judgments about life' and because of this a living literature and literary tradition can be 'an informing spirit in civilization, an informed, charged and authoritative awareness of inner human nature and human need'.[12] No one, I think, could deny that the works of Jane Austen, George Eliot, Henry James and D. H. Lawrence, of Balzac and Flaubert, of Gogol, Dostoyevsky, Turgenev and Tolstoy, of Musil and Thomas Mann are replete with charged and authoritative awareness of inner human nature and need. This awareness is brought home to us even through the manifest tensions of world-view within each writer and between one writer and another. It is just this, and because in this novelists' critical awareness and sensitivity are at a maximum, that reading them is itself a creative and judgmental process, forcing the reader to come to terms with the vision expressed and to make a judgment on it.

Music

If literature and the reading of literature clearly bear the imprint of the world in relation to man, how far might the same be true of the other arts? Take music, the most abstract and apparently least humanly involved of the arts. In a work of music – at least within the western classical tradition – a world of experienced sound is created, and within this world, through perceiving the relationships between the sounds, the listener can recognize the expression of feeling, attitude and mood, even though he may be unable to discourse explicitly on what it is he perceives and recognises. Although this fact creates difficulty for philosophical analysis, and on the level of analysis may well be something mysterious and inexplicable here, it is a fact which is part of the experience of everyone who appreciates music and responds to it with understanding. It is a fact which enabled Thomas Mann, for example, to say to Mahler that 'you express more than any other artist the spirit of our time' or for Nietzsche to castigate the music of *Tristan und Isolde* as 'unhealthy' (in contrast to the 'Mediterranean' healthiness, sharpness, rhythmicality and directness of Bizet's *Carmen*), and to dismiss Brahms as 'old-maidish'. One may, of course, disagree with these judgments, and delight in the juxtapositions of structural mastery, exquisite lyricism, biosterousness, tragedy and rhythmic strength one finds in, for example, Brahms's *Fourth Symphony*. But Nietzsche is surely right to make judgments of the sort he does: constrained by the dispositions and relationships of the notes we experience pieces of music as graceful, wistful, forceful, tragic, heroic, poignant, rumbustious, humorous, ironic, sardonic, bitter, feminine, sentimental, devout, serious, turgid, pastoral, vulgar, balanced, religious, complete, yearning, and so on and so on – that is, as evincing and bringing us directly into contact with particular emotions and attitudes, in such a way that we credit the music itself with a whole inner and sometimes even an outer life (the latter particularly perhaps in music we construe as explicitly nationalistic).

We may follow Roger Scruton here and liken our experience of a piece of music to our perceiving a smile or a gesture in a face or a body, in which something material is again seen as suffused with and expressive of human feeling and intention. In music the listener makes himself adopt for himself the feeling he perceives in the work, so as to share as from within the feeling motivating the expression or gesture. (It is significant at this point that after speaking of the way in which in art, including in music, a world is presented concretely rather than conceptually, so that art is often well understood by people *conceptually* retarded, Oliver Sacks goes on to report Helmholtz as speaking of a 'synthetic perception' for musical tones and melodies. This perception forms the basis of musical perception, and makes melodies 'faces' for the ear, recognized and felt as 'persons', a recognition involving 'warmth, emotion, personal relation'.)[13] As Scruton points out, with a musical 'face' or 'personality' we do not normally have the objective reference of the attitude or gesture, but we naturally grasp the state of mind animating the attitude or gesture: that in listening to music we have a strong sense of experiencing from within whatever outlook or feeling the gesture expresses is some vindication of Schopenhauer's claim that in music alone of the arts, we are given access to the will, to that in us which feels.[14]

What is it, though, that is expressed in a work of music? I can, of course, say *something* about this, again constrained by the musical content of the work. I might, for example, follow Glenn Gould in suggesting that in the *Goldberg Variations* there is an underlying rhythm linking all the disparate episodes, so that, though they are themselves of vastly different character, when the final reprise of the theme occurs, we feel that we have travelled everywhere and looked at everything, but also never moved from our centre: that this work gives concrete embodiment to the notion of 'the still point of the turning world'. Or I could take Deryck Cooke's eloquent account of the opening bars of Mahler's *Ninth Symphony*:

> The opening Andante, in D, has a few preludial bars stating some main ideas in ghostly orchestral colours: a halting rhythm in the bars (like a faltering heart-beat); a tolling bell-figure on harp; a sad phrase on muted horn; a rustling, or palpitation on violas. Against this background, the violins introduce the main theme, a warm singing melody redolent of summer, full of tender longing: a dark shadow has fallen across the Austrian summer which has been the permanent setting of Mahler's hours of fulfilment. The movement is a conflict between this main theme and a tormented, chromatic idea in D minor, which breaks in with bitter protest.[15]

I quote Cooke here because he seems to me to give a good sense of what the musically alert listener *hears* in those opening bars; Cooke clearly had no embarrassment about the use of such epithets as 'tolling', 'sad', 'warm', 'full of tender longing', 'tormented', and the like, and anyone who hears the music must, I think, follow Cooke in this. Nor should one jib at his referring to Mahler's Austrian summers, if, as I believe, knowledge of this sort about a work of music can amplify and enhance one's hearing of it. I do, though, have some problem with the final sentence I quote from Cooke, because it downplays the importance of the opening funereal rhythm which surely ought to pervade the whole movement and which lends it a genuinely elegiac quality, in contrast to the disjointed, hysterically charged approach favoured by some conductors; but, this aside, Cooke's verbal description does provide a set of accurate pointers to what is happening. Such commentary can be useful to someone who is as yet unable to break the musical material up into significant parts, but, no more than that, and this is a crucial point about art generally. If you want to know just what the nature of the halting rhythm is, and how it drags so ominously through the work, or just what the sadness is of the horn phrase, and how these ideas bear on the fulfilment theme and the bitter D minor protest, and so on: just what face the music has, in other words, you will simply have to listen to it. That the music has a face, and that it may or may not express the spirit of a time, or be sick and replete with morbid pathos (as the radiantly 'healthy' ear of Nietzsche would no doubt have felt) or be simply worthless (as Wittgenstein thought) is something each listener will have to judge for himself; if we take the music seriously, we certainly ought to make a judgment on it and support that judgment with some sort of reasoning and evidence. On this moral-cum-aesthetic aspect of music, Nietzsche was quite right, however much some of his own judgments of composers are wildly prejudiced and partial.

Art and self-understanding

I hope that I have said enough about the meaningfulness of music to show that music is in no sense an abstract, inhuman exercise. (Of course, no musician ever thought it was, except when in the grip of theory, or, as in the case of Stravinsky, when attempting to *épater la bourgeoisie* to boot; only misplaced theory, abstracting from experience and practice could have convinced anyone of anything so implausible.) In an obvious way, which we need hardly dwell on at length, architecture and the visual arts express and elaborate our understanding of our world and our attitudes to it. To see this in the case of architecture, one need only to compare the rococo embellishment, ornate flippancy and absence of any central focus of Empress Elizabeth Petrovna's palace at Peterhof (now Petrodvorets) with the rather over-heavy and officially monumental classicism favoured by Alexander I's architect Rossi in, for example, the General Staff building in the Winter Palace Square in Leningrad. In doing so, we will see how different were the moral and ideological worlds inhabited by the two Russian rulers, separated temporally by less than 40 years, but also by Catherine the Great and the introduction into Russia of the ideas of the European Enlightenment. In the case of painting, it will suffice to refer to the way artists are able through their work to show us new ways of perceiving the world, even at a moral and ideological level; the way, for example, Rubens in his hunting scenes suggests the tremendous fulness, energy and prodigality of nature, or the way Vermeer finds a timeless significance in a momentary domestic scene (something for which he was much esteemed by Proust), or in Claude's nostalgic evocations of a never actually realised classical world, or in Francis Bacon's depiction of human beings as skewered carcasses and writhing meat.

There is more to be said on the implications of all this for both art and morality. For the present, though, it is enough to have established the manner in which works of art can, if they communicate successfully, capture, give expression to and develop our ways of experiencing and evaluating the world and, in so doing, contribute to our understanding of ourselves. They do this in a concrete way, through particular experience, by reference to the concrete particularity of things. They evoke possible or actual reactions to things and present the audience with these reactions for their contemplation. In this way, works of art bring us face to face with the particular ways in which things affect us, and are, as Goethe rightly said, the imprint of the relation of the world to man. They are not an abstract, external, formal account of that relation, but take the perceiver of the art work through his perception of the work, into the subjective visions others have or could have. The type of understanding of ourselves and our world which works of art make possible is quite different from that made possible by natural science. Indeed, even if the nightmare of a complete science of human experience and behaviour became a reality, in so far as human beings continued to have subjective experiences of their world, there would still be a need in whatever culture remained for works of art in which human responses to that predicament were expressed. And even within an all-embracing religious framework, the objective account which the dogma propounded of man's relation to his God and the rest of the world, would not simply leave room for, but cry out for expression of the subjective experience of that relationship. It is hardly surprising that much of the most sublime art of our culture has been just that, from Dante and early Italian masters and the medieval cathedrals to its final magnificent, but anachronistic flowering in the last three symphonies of Anton Bruckner. But the age of a general unquestioning Christian faith had passed long before Bruckner. With it, most artists lost the confidence in belief necessary to produce such works. Even in professedly religious artists and writers of the nineteenth and twentieth century there is generally a more self-conscious, questioning spirit – a consciousness that the world and God are in tension if not in flat-contradiction – than in Bach, say, or Fra Angelico. Dostoyevsky is a typical figure here, with the all-pervading sense in his work that in asserting religious values one is taking a stand against the world.

In works of art and literature, then, we come to a particular, inward type of understanding of human life and experience. This inwardness derives from the way in which in a work of art there is a sympathetic enactment of the experience depicted or evoked, and in which, on the part of author and audience, feeling and receptivity are at their peak. In a great work of art a type of understanding of ourselves and our predicament is made available to us through the experience and contemplation of such works which is not available either in less perceptive, more fantasy-ridden and stereotyped works (with which current 'popular' culture is largely populated), or in science. But it might be felt that the so-called social sciences supply in a possibly superior, because not personal and idiosyncratic way just the sort of understanding of our condition which I have been claiming comes to us primarily through the arts. Thus, for example, Bernard Williams apparently wants to 'associate' the humanities (by which he means history, philosophy, and the study of literature and the arts) with 'certain areas of the social sciences', and to defend the study of the humanities by claiming that they are in the same business of social understanding and criticism as the social sciences.[16] In the first place, this is a very restricted account of the role of the arts in our apprehension of ourselves and our lives and the possibilities for our lives. We may have been so brainwashed by social critics that it is sometimes difficult to remember that most of our lives are lived on a much smaller scale than anything that could be dignified with the name of 'social' or 'political'. Ordinary people have to live their lives on an ordinary, personal level; there are decencies and betrayals, judgments and perceptions which persist above or below whatever political circumstances exist or whatever social changes come into force, and these are the stuff of the ordinary moral life. Our English literature from Jane Austen to Iris Murdoch and Antonia Byatt has a fine tradition of reminding us of such things, and of teaching us the ordinary, and because so ordinary, easily overlooked moral nuances that suffuse the most everyday encounters and relationships and which wound or bless so much. Precisely because the everyday moral world is a matter of nuance and perception, rather than of grand gesture, art in general and literature in particular is a better way of demonstrating what it is to lead a moral life (or an immoral one) than philosophical theory, which is in its generalising will inevitably be largely blind to the small and important.

Art and social science

But over and above criticising Williams' restrictive view of the humanities as important because of their role in *social* criticism, it is important to compare the types of understanding of ourselves that might be given to us in the social sciences from that which the arts can give us. If the social sciences are thought of as consisting in the statistical tabulation and analysis of data – a perfectly proper and useful activity if, for example, we want to know the effects of a given policy – they will prescind from the experiential impact things have on agents. Reports of this sort could be compiled by Martians or computers: or to put it another way, you will not find out what prison life is like and the precise nature of the moral corruption it seems to generate by reading journals of criminology, but you might by reading G. F. Newman. The objective sociological report lacks just that attempt to convey the subjective experience of its subjects which might present one with a motive for action. And the same is true of the type of analysis of social and economic orders inspired by the example of Adam Smith. Here the aim is to show how a social system can arise quite independently of the intentions of participants in it, how, for example the self-interested acts of people engaging in market activity can produce a system which might promote the well-being of all. Such analyses are an essential complement to attempts to capture and describe the meanings and values of the agents in a given society, but they cannot supplant them. After all, it is a feature of a particular society that its agents are largely motivated by 'selfish' market considerations, and one that is not found in all societies.

Many workers in the social sciences do attempt analysis and evocation of the experience of agents in a particular culture or situation. Much of the best anthropology can be seen as attempting to recreate or order the experience of a particular tribe or stratum of society. But at this point the sociological or anthropological narrative comes closer to what I am seeing as literature (and the same goes for much good historical writing). The reason for this is that while such narratives will be constrained by the facts observed by their authors, they are, in essence, imaginative reconstructions of a *Lebenswelt* or part of a *Lebenswelt*, and criticisable as a novel would be, in terms of the accounts given of the meaning and experiences of the subjects rather than anything confirmable by empirical, scientific, objective methods.

It is important to emphasise, though, that although the description of the experience of agents in a society is not a scientific study and cannot proceed by strictly scientific methods, it will not be confined to what agents are immediately conscious of or explicitly intend. The tone and morality of a society may well be deeply conditioned by linguistic and other practices which are so pervasive that they escape notice by members of the society, and form only the background to their explicit intentions. We could think here of the way different societies have defined the scope of the notion of a human being, and whether infants, slaves or mental defectives have been included in it, and the way such definitions might bear on the life and motives of a people. A social scientist or anthropologist, coming from outside with different assumptions, may be in a good position to pick out and notice such unconscious assumptions. But here again the skill is essentially the same as that of a good novelist, bringing to light unnoticed presuppositions in the behaviour of a character. Whether we accept the accounts in either case will depend on the extent to which by crediting a character or a people with the postulated unconscious meanings and intentions we make their behaviour and experience appear more intelligible and the fabric of their lives more coherent. Such descriptions and evaluations remain on the level of meaning and experience, are not like the postulation of unobserved particles in physics, whose significance is given completely in terms of the effects they are held to have and does not imply anything about the world perceived or experienced by those influenced by the effects.

There remain to consider the full-bloodied ideologies which have been produced within the social sciences, such as Marxism and psycho-analysis. I cannot go into any detail here, but in view of the manifest conceptual and empirical difficulties faced by these theories,[17] there is certainly a large body of opinion, even opinion favourable to Marxism and psycho-analysis, that what we have here are not really sciences at all, but moral visions expressed *modo scientifico*. If this is so, then what we have in each case is something akin to artistic perception, in which the world is presented by the artist as having a particular impact on the perceiver. And if this is so, then as with any postulation of unconscious intentions or motives the 'yes, but' of literary criticism might well be the right way to consider these theories. What is being overlooked by seeing our lives in Marxist or Freudian terms? Yes, you can see human relations in a bourgeois society in terms of class struggle, *but* does this exhaust the reality of working life? Yes, dependence on the state for reward might seem attractive at first sight, *but* how far would such a thing distort normal human expectations. Yes, everything can be seen as a sexual symbol, *but* is doing this itself the disease we suffer from, rather than the cure (as Karl Kraus suggested)? Yes, you can look at sexual desire as a hydraulic system needing periodic release in any way available, *but* how far does this distort the directedness to a particular person of human sexual desire? What I am suggesting is that the grand theories of the social sciences may be more like a novelist's vision in adumbrating a particular possible subjective perspective on the world than anything properly scientific, and that, as such they are to be handled not as sciences (in terms of their predictive successes or failures), but in terms of their adequacy in accounting for the meanings we find and the experiences we have in the world of human meanings and experiences.

CHAPTER 5

Truth and Art

> He knew at that moment he was withdrawing from everything, not merely from human beings. A moment more and everything will have lost its meaning, and that table and the cup, and the chair to which he clings, all the near and commonplace things around him, will have come unintelligible, strange and burdensome. So he sat there and waited until it should have happened. And he defended himself no longer.
>
> (R. M. Rilke, *The Notebook of Malte Laurids Brigge*, London, Chatto & Windus, 1930, p. 49)

Rilke's vision is of a dying man. A ghost, occupying his own place at his own table, a *Doppelgänger* of the one who is alive, he can no longer recognise the world as charged with human meaning. As for himself, terrified by the vision of the other who is dying, for whom the world has become so unintelligible, strange and burdensome, 'I should so gladly stay among the significances that have become dear to me; and if something must already change, I should like at least to be allowed to live among dogs, who possess a world akin to our own and the same things.'

Would the scientific ideal of an absolute conception of the world, assuming such a thing could be delivered, be akin to a death of the human spirit? As Rilke describes death, it clearly would be; in the absolute conception of the world, aptly characterised by Thomas Nagel as 'the view from nowhere', things and the world in general would lose their significance for us. A view from nowhere would not have the world centering on the conception one particular consciousness or species has of the world, and things as, in the first instance, relating to it, and to its manner of perceiving and intervening in the world. But, as we saw in the chapters on science, there are always going to be limits on the extent to which we, as embodied beings, with a very particular and specific perspective on the world, can transcend that perspective for a more objective view. We will not be able to assert the truth of any conception of the world, except to the extent that it can be verified in and through our experience of the world. Beyond that, the conception remains speculative and, for all the evidence we have, only one of many possible conceptions of the world. The conception, too, that we produce, for all the distance it takes us from the everyday, will still be framed in terms of models intelligible to the human mind, at the risk of itself being incoherent for us. The example of quantum physics is, of course, instructive here, with its paradoxes and conundrums suggesting that that particular level of reality is likely to remain resistant to understanding by us. Furthermore, there seem to be difficulties in giving an objective, decentred account of our conscious experience; from the absolute point of view, the phenomenon of consciousness remains mysterious. And physics itself leads us to think of our experience of time as irreducibly relative to our particular perspective on the universe. But even if we framed for ourselves a theory which regarded time as relativised to space in a space-time manifold, in which we represented ourselves as space-time worms embedded in the space-time continuum, in so far as we contemplated and worked on the theory, we would once more take up a temporal stance to that product of our consciousness. Even in contemplating the scientific unreality of time, the centrality of time to our experience re-asserts itself, in which the succession of moments is real enough given that our experiences are uncombinable and necessarily non-simultaneous.[1]

So, for a number of reasons, we cannot avoid or eliminate human perspectives on the world, even as we try to transcend them. The lesson to be drawn from our consideration of science is that our conceptions of the world will always be tied down to and rooted in the ways things affect us and the ways we deal with them. Reality can be disclosed to us only in and through our viewpoints and concerns. For conscious beings in the world, the touchstone of reality is in the end how reality is revealed to them and their particular form of experience.

The literal and the metaphorical

At the level at which Rilke contemplates the other as he leaves the world, it is obviously a human world he is leaving. Tables, cups and chairs are human artefacts, being the things they are because of their functions in our lives. And even our concepts of things we regard as existing naturally in the world apart from human intervention are at least initially given to us in terms of the appearances things have to us and their roles in our lives, although, as the example of the ornamental marbles shows, we may later revise our original classifications of naturally existing things in the light of scientific discoveries. The world in which we live our daily lives, the human world, is very much a world whose classifications and concepts can be seen as corresponding in a broad sense to human interests in the world, though there is, of course, a mutual reinforcement and dialectic here, for the interests of any given group of human beings will be very largely determined by the concepts and classifications given to it by the language and culture it inherits.

Saying that our concepts and classifications relate to and form human interests is not, however, to say that the statements we make about the world are not true or false as the world determines. Whether the statement 'My dog is under the table' is true or false is determined by the world, not by human beings. But that we should be classifying and thinking of the world in that sort of way is a function of our imposing on the world the language and concepts we have, which again is a function of our human perspectives and interests. Even our literal talk about the world, then, in a deep sense presupposes and reflects a specifically human viewpoint on things. The similarities in things picked out by our concepts reflect features of those things which at some level or other have seemed important to human beings; naturally it is at the level of everyday social activity that this interest – relativity of concepts and classifications is most evidently based in the ways things initially appear to us and affect us.

The everyday world – of cups, tables, chairs and dogs – is already a world transformed by human imagination. Because it is the everyday world, that which we enter in our first learning our language, that it is a humanly transformed world is easily overlooked; perhaps we can imagine a mythical time at which the first groupings of, say, the diverse creatures we now all see as horses might have seemed utterly fantastic and non-literal to the mass of humans, something which violated the previous well-embedded systems of classification and appearance. This thought is the root of the conception to be found in Vico and subsequent romantic thinkers such as Shelley and Herder that all thought and language has its roots in poetic leaps of imagination, by which the entities that science will later define and dissect were originally created. This position was later embraced, notoriously, by Nietzsche:

> What, therefore, is truth? A mobile army of metaphors, metonymies, anthropomorphisms: in short a sum of human relations which became poetically and rhetorically intensified, metamorphosed, adorned, and after long usage, seems to a nation fixed, canonic and binding; truths are illusions of which one has forgotten that they *are* illusions; worn out metaphors which have become powerless to affect the senses; coins which have their obverse effaced and now are no longer of account as coins, but merely as metal.[2]

Nietzsche's thesis is clearly somewhat more than a simple re-statement of the point already made, that the use of general terms or concepts requires that we fix on some interest-determined similarities in a set of things, in virtue of which we classify them as being of the same kind. For it would not follow from that that all truth is an illusion or that all language is initially metaphorical. Nor indeed could any such theses be true for any activity we could recognise as the speaking of a language at any stage of its development. For the interpretation of the linguistic utterances of others involves our ascribing to them beliefs and other attitudes on the basis of their utterances and predicting actions on the basis of this. But *which* belief should we ascribe to Gaston when he asserts, say, 'La neige est blanche'? Presumably that snow is white. But in order to conclude that, we have to assume that Gaston is using his words with their standard or canonical senses; viz, 'La neige est blanche' means that snow is white. Only indeed on the assumption that words in a language have standard or canonical senses will we be in a position to ascribe relevant beliefs and attitudes and predict actions to speakers of that language, and the same goes for our own use of our own language in our attempting to convey to other English speakers our own beliefs and attitudes.

Now, someone interpreting or understanding the assertions of others will in the normal case assume not only that the words used are used with their standard senses – or there could be no pattern of interpretation or consistent understanding at all – but also that those others are intending to express their beliefs by their utterances. People who lie indiscriminately will not succeed in communicating to others because their audience will never know whether they are to be taken as telling the truth or not. Equally people who use words in their own idiosyncratic ways will not succeed in communicating because the audience will not know what it is they are intending to convey. But it is also a feature of language use that in addition to inveterate liars and people whose grip on the regularities of sense in a language is insecure, we and others often use words with their standard meanings, in what appear to be assertions, to express propositions which we or they could not conceivably be regarded as believing straightforwardly. For example, suppose I say not that snow is white but that my conscience is white. Now, given that I am using 'conscience' and 'white' canonically and that I am minimally rational, I obviously cannot believe that my conscience is white because my conscience is not a material object which can possess a colour of any sort. But because whiteness has certain well-known associations (of purity, innocence and the like), my listeners will probably interpret what I am saying to the effect that I am innocent in some relevant sense, but this shift from the literal to the figurative *via* the associations of whiteness is just what constitutes a metaphor, a *removal* of sense. Hence, for the metaphorical to be possible at all, there must be a literal sense of the words involved, in which it is possible for literal truths and falsehoods to be expressed. It is through the failure of the literal interpretation of an utterance to be a plausible interpretation of what a speaker is saying that an audience recognizes that an utterance is intended metaphorically. Against Vico and Nietzsche it follows that the whole of a language, even at its start, cannot be metaphorical, however much any linguistic grouping of distinct objects requires or required leaps of imagination to focus on the similarities between those objects and to disregard the differences between them.[3]

Why, though, should we be concerned to engage in metaphorical talk, subverting the literal? Literal talk is tailored to speaking about those things of which we have direct experience, and which are publicly accessible. It is the public accessibility of the things that we talk about in literal terms that makes language learning possible. The connection between words and such things can be communicated to learners of gestures and signs; on this basis, learners can pick up rules governing the use of words, and by further successes or failures in using their words in achieving the results they want in the world, come to have their understanding of the rules of the language reinforced or corrected. In other words using words in conformity with public, socially understood standards of use depends on those words having connections in their usage with publicly accessible things and events, and this is the basic presupposition of any possible language.

Metaphor in science

The users of a language will, though, wish to use language in a number of ways which do not consist in simply referring to publicly accessible objects or events. Language is undoubtedly extended beyond primitive literalness in scientific enquiry, where it is assumed that we live in a world, much of which is inaccessible to direct human observation. Words will be introduced in science to refer to entities which we cannot directly observe, and the proof of whose existence is indirect, and through the observable effects of the postulated entities. It is natural to interpret the terms we use to refer to such entities in a literal way: to think of 'meson' or 'boson' as referring to objectively existing entities, just as much as 'tree' or 'dog'. Despite the difficulties in establishing the truth of theories referring to the elementary particles of physics, there is something persuasive and plausible in going for a literal interpretation of such terms. Part of our sense that we live in an objective world much of which we do not have direct access to is to admit that there are things and events beyond our ken. Even though we cannot establish directly that there are mesons or bosons or not, and may never be in a position to do so, it is surely right to regard any of our statements about mesons and bosons as being true just in case there are mesons and bosons doing whatever it is the statement asserts of them, and false otherwise.

Even though in science we ought to interpret its statements about things and events literally, much as we interpret statements about everyday things and events, despite the indirectness and uncertainty of our knowledge here at times, there may well be metaphorical aspects to our subjective understanding of the statements of science. What are mesons or bosons like? To answer this question, we may well appeal to aspects of accessible, public experience, just as has famously been done in thinking about gas molecules as elastic balls or atoms as little solar systems. Thus, bosons are 'eaten' or 'absorbed' by other particles, mesons 'knock' electrons off atoms. To speak, correctly, of such thinking as metaphorical is to assert that we know that gas molecules and atoms, bosons and mesons are not *exactly* like that, and that we do not intend the comparisons to be taken literally. Undoubtedly metaphorical thinking of this sort plays a central role in the formulation of scientific thought, precisely because we do not have direct observational access to the entities and processes in question. Metaphors are the most powerful instrument at our command to drive beyond what can be established by observation of the directly accessible. But in science it becomes important to make precise the limits of the metaphor, to show just the respects in which an atom, say, is like and unlike a solar system. And once this is done, it is unclear how much of the metaphor actually remains, except as a device for introducing what are now precisely clarified ideas to a beginner. Moreover, the value of a scientific metaphor will always in the end be determined by its effectiveness in leading to literal truths in the objective and publicly accessible world: the predictions and experiments by which we test any scientific theory and weigh its practical value.

Metaphor in religion

The scientific is not the only realm of discourse in which we attempt to transcend the limits of the observable. In religion, too, we attempt to describe realities to which we have no direct experiential access, and here metaphorical modes of discourse seem rather more indispensable than they are in science. It is not just that religious myths, unlike the scientific, do not have direct and precise predictable consequences. It is rather that, as Wittgenstein stressed in his reported remarks on religious belief, in religion 'the whole *weight* may be in the picture'.[4] He contrasts the case of a representation in which the precise mode of representation is immaterial, where one picture might just as well be used as another, one projection of an ellipse or some other, chessmen of one shape or another, with the case of a religious image, such as Michelangelo's *Creation of Adam*. Here Wittgenstein says 'there is nothing which explains the meanings of words as well as a picture.'[5] We know, of course, that God is not an old man wrapped up in a queer blanket, and we know that Michelangelo did not intend us to think he was. God, for us and for Michelangelo, is, if anything at all, a disembodied spirit. Yet everyone who sees that pictures feels, as Wittgenstein did, that this picture is as good a representation of the Creation of Adam as anything could be. Unlike a scientific model or metaphor, though, it is not a dispensable metaphor, nor is there any way of extracting the literal content from the pictorial or metaphorical. Someone who attempted to distil the essence of a religion from what are recognized by all involved to be metaphors, pictures and parables, would not only fail in the sense of producing only a projection of his own feelings about the religion – as the sorry story of the nineteenth and twentieth century attempts to demythologise Christianity have all too clearly shown – he would also have misunderstood the phenomenon of religion.

Religion is essentially an attempt to see the universe in personal terms, in terms of human personality. For the religious, the universe has a face, or is the face of some utterly Other being beyond. This face the universe has or is is not taken by the believer to be an empirical fact. Although it may for some manifest itself in quasi-empirical revelations such as miracles, what is really at issue in religion is true nature of reality itself, beyond and behind *all* empirical facts or scientific explanations. The question is precisely whether the universe – including all empirical facts and scientific explanations – is to be seen as the manifestation of a will, an intelligence, a person even, a being, at least, whose inner essence in some way conforms to our own.

This 'personal' interpretation of the deep nature of reality is undoubtedly hard to reconcile with another crucial aspect of the religious drive, the assertion of the absolute transcendence and otherness of God. For only a being utterly transcendent and other could provide the cessation of all questioning and explanation which religious seekers are looking for. Further questions inevitably will be provoked by more assertions of a determinate sort, such as 'The world is as it is because God so willed it'. Why, one will ask, did God so will it? I have argued at length elsewhere that these two aspects of the religious quest are in a mutual tension which is ultimately destructive of religion itself.[6] Whether this is so or not, for our present purposes we need only reflect on the ways in which in a religion the universe itself is seen in personal terms.

Two types of religion may appear to provide counter-examples to my general thesis here. First, there are primitive or so-called primitive religions in which the world is the abode of many gods and spirits, but not completely under their control and not under the control of any single divine force. I do not see this type of religion as a complete counter-example to my general thesis about the personalizing view of the universe implicit in religion. According to Walter Otto, the classical Greek 'at every turn of life saw the visage of a god'.[7] Apart from the cosmogonic myths of Olympian and other polytheistic religions in which the many gods are taken to have a common parentage, it is only a short step in thought from seeing the world as full of many anthropomorphic gods and their manifestations to seeing it as emanating from one transcendent source, a step made, of course, during the evolution of Hinduism.

Buddhism is perhaps a more telling counter-example to the thesis that religion involves seeing the universe in personal terms, with its apparent denial of any transcendent being, and the Buddha's own constant 'not-this, not-that' and refusal to engage in theological discourse or speculation. Despite this, it is surely reasonable to see the religion of the Buddha as an attempt to maintain the attitude of mind inherent in Hinduism. In Hinduism, human life and human consciousness are supremely important because of their participation in the absolutely Other life and consciousness underlying the universe itself, a participation whose fulfilment requires the extinction of all separate identity and separateness of mind on our past. Buddhism can be seen as an attempt to recapture the spirit of all this without engaging in the conceptual difficulties involved or being enmeshed in what would be seen as the 'nonsense' of the Vedas. How far, though, one could go in understanding what it was that the denials of Buddhism signify and what its spirit is without the myths and mythology of the Vedas (or some more specifically Buddhist mythology) is extremely doubtful, and is part of the reason why Wittgenstein is right to emphasize the importance to religion of a mythology. The mythology cannot be factored out as in science, while a religious *via negativa* even in as extreme a form as that of the Buddha or Karl Barth (with his insistence on the absolute otherness of God), provides a route to enlightenment only on the back of some *via positiva* (or *metaphorica*). The denials have force only given a sense of what it is that is being denied, in order, as it were to extract the indeterminate spirit from the all too determinate mythology.

Aside from its relationship to its Hindu roots, it is clear enough that Buddhism, even in its most austere and negative forms, does propound an intimate relationship between man and the cosmos, in which one is responsive to the other, and one reflects the inner essence of the other. There is all the difference in the world between Buddhism and the atheism of a Bertrand Russell, which is based on the insensitivity of the cosmos to man and man's chance-based and cosmologically insignificant life. For Buddhism, as for other truly religious systems of thought and practice (in contrast to materialistic cults such as Marxism), there is a sense that man's deepest and truest personal aspirations are somehow a reflection of the inner essence of or behind the universe, and that the universe or what is behind it are not insensitive to the humanly significant. It is for this reason that religious belief is most naturally and fully expressed through metaphors which represent the divine in human form, imbued with humanly significant responses and attitudes. How else, indeed, could the universe or its source be seen or experienced in terms of personality?

Metaphor seems particularly appropriate for the elaboration of religious thought, because in religious thought one is suggesting that the greater than human has characteristics analogous to the human. As such, of course, a religious picture or metaphor will, if subscribed to, evoke responses and attitudes from us. To see God as Michelangelo's Creator, or as the father of the Prodigal Son is already to take up certain attitudes to God. To see the world as created by Michelangelo's Creator or by the father of the prodigal son is to respond to the world in a particular way.

It may, though, be asked, in which way? What I now want to suggest is that apart from the other reason as already given, it is no accident that religion needs metaphor; nor is it an accident that metaphors, broadly conceived, play a significant role in the arts. For the adoption and elaboration of suitable metaphors has a significant and central role to play in the expression, elaboration and concretisation of feeling, experience and human attitudes to the world.

Metaphor and the expression of inner attitudes

Literal discourse, I have argued, basically depends on shared access to the publicly observable in order to fix standard meanings of terms.

Saying this does not mean that we cannot refer to the inner attitudes of others, for inner feelings, experiences and attitudes such as pain, grief, happiness and joy have outward manifestations in behaviour and action. We know the circumstances likely to evoke such feelings and attitudes on the part of others, and we know how people normally express themselves – in words, behaviour and actions – when they are happy or grief stricken or in pain. In this sense, feelings, experiences, and attitudes on the part of others are publicly observable by us and by others, and our own are to others. It is no mystery, then, how a public language referring to feelings and attitudes is possible.

But feelings, experiences and attitudes are not just public phenomena in the public world. They are also felt and known from inside, by the person subject to them. A literal description of a feeling or attitude I have will not precisely delineate it, nor will it bring out the way in which it is not an object for me, but something I feel, something constitutive of what I am. It is at this point that one can have recourse to metaphor or symbol, transferring certain terms from the public realm to indicate the nature of one's inner state. If one describes an experience of dread, say, as like a mist closing in on one, or a grief as too heavy to bear, one does not imply that one's mind is enveloped in mist or that there actually is a physical weight bearing down on one. But in using such forms of description one may convey to one's audience both the nature of the feeling or attitude, and, precisely because of the cancelling of literal sense in the metaphorical usage, that one is not describing a public object (or, indeed, any sort of object), but indicating one's inner state.[8] And the metaphor, precisely because it is not literal, awakens, intimations and a free flow of associations, where the literal closes and confines one's thought.

It has been suggested, by Hayek, that the language we use to describe our sensory states is not metaphorical; according to Hayek because we all tend to agree on describing colours as having a soft or sweet appearance, tones as thin or dark, tastes as hot or sharp, smells as dry or sweet, and so on, as well as on the emotional characteristics of sensory qualities, what is at issue here is a genuine sensory order, with a unitary character and its own types of differentiation.[9] Part of what Hayek says is true, and striking, although I am not clear as to whether softness and sweetness, thinness or darkness, are on the same scales of comparison at all. But it is true that we do tend to agree on the types of experience and emotions specific sensory qualities arouse in us. On the other hand, the descriptive epithets Hayek instances are clearly metaphorical as applied to experience. In some cases, as with the cutting pain, the origins of the epithets may have to do with the sorts of feelings and experience things of that type produce in us. A cutting pain may be the sort of pain knives produce in us, and a dry smell may be the sort of smell dry things characteristically have, but transferring the dryness and cuttingness from the objects that are dry or cut to the experiences we have is clearly a metaphorical device. And I do not think that any such explanation could account for sweet colours or dark tones.

What is the criterion of success for metaphorical usages in non-scientific cases where there can be no question of the metaphor standing proxy for a more literal description? It is clear enough that part of the aim of metaphor in both religious and experiential-cum-attitudinal cases is to get one's audience to share one's experience or attitude, or at least see what it might be like to have such an experience or attitude, as well as making precise to oneself what one's experience or attitude is. So the criterion of success will be to produce a metaphor which evokes the right sort of experience in one's audience. The success of a metaphor is thus to be distinguished from that of a literal or scientific statement which will be taken to be true if it corresponds to the way the world is, independently of how anyone feels about it or responds to its imagery.

Homeric metaphor

Some metaphors are in practice dead, by being so worn that everyone has forgotten their metaphorical character and treats them as if they were literal epithets. But metaphors which are not dead in this way will demand some sort of active engagement on the part of the audience, for in a metaphor the audience is being asked to experience the world in a particular way, and this is an open-ended active process, requiring the exercise of one's imagination and the engagement of one's feelings. Opening the Iliad at random, I read of Agamemnon rampaging through the Trojan ranks:

> others still in the middle plain stampeded like cattle
> when a lion, coming upon them in the dim night, has terrified

> the whole herd, while for a single one sheer death is emerging.
> First the lion breaks her neck caught fast in the strong teeth,
> then gulps down the blood and all the guts that are inward;
> so Atreus' son, powerful Agamemnon, went after them
> killing even the last of the men; and they fled in terror.[10]

Homer is asking us to look at Agamemnon among the Trojans as if he were a lion preying on a herd of cattle, the herd in panic and its slowest member savaged by the lion's teeth, and its blood and guts swallowed by the lion. To do what we are invited to do here we have to transcend a literal account of the incident, looking at it in a new light: a light which allows us to experience the scene as horrific and terribly exhilarating at the same time; and also and above all (like so much else in Homer) as part of the natural course of things, Agamemnon as a lion among cattle, naturally a king, among inferior animals, ill-equipped for fighting. The connections and comparisons Homer suggests encapsulate an attitude to this incident and by extension to heroic warfare in general, an attitude which, of course, captivated and formed part of the inner life of many of Homer's more devoted readers from Alexander to Nietzsche. And even while being exhilarated by the sheer power and beauty and triumphs of the lion's assault – and so entering into the experience of Agamemnon himself and also, by contrast, into the terror of the Trojans – we might remember that in classical times lions were regarded as wild and dangerous and a threat to farmers and shepherds, as far as possible to be removed from the land. In sharing in the exhilaration of seeing Agamemnon as a lion and, in Agamemnon's own exhilaration, are we also to embrace this aspect of the comparison? Or is it a dispensable part of the metaphor, like the lion having four legs? The lack of determinancy in answering such a question is what I mean by the open-endedness of metaphor. A good poetic metaphor is one that opens up a field of association and meaning which, precisely because literal meanings are suspended, is clear enough in its central sense but unbounded in possible further resonance.

In distinguishing metaphorical from literal speech, I have attempted to show how metaphorical speech is particularly appropriate for reaching out to the not directly describable. In the case of scientific description of the not directly observable, metaphor is treated as dispensable. Ideally, we feel, literal forms of description would be what we want. But with talk of feelings, experiences and attitudes – discourse about the inner life – direct forms of description can at most give us the outer shell of the inner. Just what was Agamemnon's sense of exhilaration like? Just what does the heroic attitude to battle consist in? Homer's metaphor of the lion and the cattle gives partial answers to these questions. But they are answers which will do their work only for those who know how lions prey on herds of grazing animals, who can sense something of the excitement and terror of such a scene, and who are capable of feeling the point of the comparison when the image is transferred to men in battle. I say feeling rather than understanding here, because the point of a poetic metaphor is to convey a possible experience or attitude one might have as a human being to some situation, and to see that situation as charged with human significance, rather than to give a merely theoretical conception of it. Just what conceptions of God are conveyed by Michelangelo's Creator or the parable of the Prodigal Son? It is clear that like metaphor in art, these images do not convey theoretical conceptions of the divine, but through the image convey simultaneously a conception of God and an attitude to God; and both will be conveyed to those who are capable of responding to the images, but in no other way. As with a metaphor in an artistic context, as Wittgenstein said, the whole weight is in the picture.

It would be rash to claim that all artistic expression depends on metaphor in any straightforward way. It is not easy to see how the notion of metaphor can be given any precise or central role in music, architecture, in many works of painting or sculpture or even in the realistic novel. None the less, I now want to suggest that all artistic expression shares certain features of metaphorical modes of discourse, as a result of which our response to a work of art will be quite different from our response to a literal description of a state of affairs or a scientific theory.

Art and imaginative transformation

The first thing to be said about works of art is that even where their medium is realistic, as in certain novels or certain genres of painting, such as landscape or portraiture, the aim of the activity is not simply to reproduce the world literally or photographically. While there have been plays and even sculptures which have apparently attempted to reproduce an actual conversation or a particular room exactly, unless the artist is attempting to make some sort of socio-political point (and perhaps even if he is), such works fails as works of art because of the intentional absence on the part of their creator of any imaginative insight and his unwillingness to present anything as transformed by his own imagination. It is hard, indeed, to distinguish them from journalism, a simple piling up of ephemeral facts and reproductions of facts. Perhaps part of the problem one experiences even in reading a work like Flaubert's *Salammbô* arises because in it Flaubert piles on meaningless fact after meaningless fact. But his attempt to create a verisimilitudinous account of a perfectly inhuman world of cruelty and excess is vain and in this curiously dead work actually succeeds only in convincing the reader of his inability to overcome his sense of *taedium vitae* even by the most exotic and extreme of means.

The *Iliad* describes events every bit as horrible as those in *Salammbô*, but is vibrant and full of life just where *Salammbô* is dead. In the *Iliad*, as we have seen, descriptions of battles are characteristically enlivened by metaphor and presented in terms of full human significance. Few works of art approach the richness of the *Iliad*. Nevertheless from the *Iliad* we can extract a central feature of any work of art, which is what distinguishes a work of art from a journalistic record. In a work of art, there is a sustained attempt to endow aspects of the world with human value and significance and to present us with ways in which those aspects of the world might be experienced. In the painting of landscape an artist may, for example, emphasize the way in which the rising sun can caress and enliven limestone cliffs, so as to bring a golden glow from them, or the way in which the blue of the sky, the tumult of the clouds and the white of the snow complement each other, so as to create an impression of mountainous peaks as pristine waves in the sky. Hackneyed as such impressions may now be, they were quite new and fresh in the early sixteenth century when they were first brought out by Albrecht Altdorfer in his extraordinary *Battle of Arbela* (Munich, Alte Pinakothek). The example brings out well the way in which even a landscape painting will not be a merely passive, mechanical record of a scene. (One could say, photographic here; but that overlooks the way photographers manipulate filters and lenses to create just such effects.)

Landscape painting, then, will aim through emphasis and selection of detail to concretize particular ways of experiencing a scene, and will in some cases actually succeed in transforming the experiences one has of a type of scene; and the same goes for other types of pictorial or sculptural representation. Much the same will be true, albeit in a different medium, with realistic plays and novels. The selection of detail will always be to present character and plot to be experienced in specific ways, and we will read the painting or novel or play correctly only if we understand the way we are intended to respond to the characters and the plot. Although metaphors may not be used directly, the process of composition and interpretation is somewhat akin to the use of metaphor, in that in both types of cases particular features of a scene or event are fixed on in order that the audience might experience that scene or event in a particular way. Success in both cases is, in the first instance to be judged not in terms of literal truth of representation, but in terms of success in transmitting a vision of the world as experienced or experienceable in a specific way. (This does not mean, of course, that we actually have to agree with or share the vision. When Oscar Wilde quipped that that man must have a heart of stone, who, on reading of the death of Little Nell, does not immediately burst out laughing, he certainly understood Dickens' intentions.)

In the cases of music and architecture we do not normally have either straightforward metaphor or representations of the world. Nor can the aim be plausibly construed as the presentation of the world (or even a bit of it) as experienced in a particular way. On the other hand, in both music and in architecture, in so far as the latter is regarded as an aesthetic form, the aim could be said to be the having by the hearer or onlookers of certain specific experiences brought about by his contemplation of the work. And these experiences will typically involve his 'reading' the work in a non-literal or even metaphorical way. In contemplating a classical building we will see a cornice modifying the thrust of a perpendicular column by emphasizing the strength and stability of what lies below, and anticipating the roof or dome above; we will see the capital of a column bearing the weight of the pediment above; a dome floating above the building; the columns of a portico and figures on a pediment in rhythmical terms; and the whole building in terms of balance, sobriety, grace, and dignity, as the case may be.

John Haldane has suggested, plausibly enough, that in so far as we see architecture in terms of human needs, intentions and creative activity, we naturally tend to 'relate the articulation of compositional elements and its resultant effects to aspects of human nature', and think of buildings in terms of their straining, stretching, pointing, striding, reclining, being upright, cast down, stately, confident, at ease, erect, attentive – and, on top of these characteristics, as secure, stable, graceful, majestic and so on.[11] Architecture, then, invites us to see a building not just in terms of a human face, but in terms of a whole physique, demeanour and character. Fanciful as this may seem to some, there is certainly a natural tendency in looking at buildings (and indeed at other human artefacts) to transfer epithets initially appropriate to the domain of the creator to the domain of the objects created; it is in these terms, indeed, that aesthetic appreciation reflects the moral stance of the viewer and the human quality of the object viewed.

In listening to a piece of music, the listener construes the literally static tones he hears, which are the tones they are only by having a single pitch, as embodying movement, flowing into each other in melody and also as stratified, as high or low as if they occupied positions in space. Similarly in perceiving the rhythm of a piece of music, we hear something over and above mere temporal succession of sounds. As Roger Scruton puts it, 'we hear sounds joining to and diverging from each other, exerting over one another peculiar "fields of force", determining each other in a manner familiar from our knowledge of human movement.' In hearing sounds as music, what we hear is thus interpreted in terms of space and movement and rhythm; these aspects of the experience are brought to it by us, and only thus do we hear music as opposed to meaningless sound.[12] In order to understand the music of a culture, one has to be able to do far more than this, of course; it will be necessary to experience particular tones as having qualities such as thinness or darkness (to use Hayek's example), and particular phrases as yearning, or noble, or elegant. We will ultimately have to have the ability to hear, as Thomas Mann did, the E flat major triad of the deep flowing Rhine at the start of *Rheingold*, as expressive of the origin of the saga, of music and of the world itself, and eventually, no doubt, to build up an understanding of the experience of a whole work from one's understanding of its elements and their relationships.

While music and architecture do not present us with metaphors, what little I have said in the previous paragraph about the understanding of music and the aesthetics of architecture suggests strongly that we need to experience works of music and architecture in terms which may be regarded as metaphorical, in that they go beyond the types of property which stones and sounds literally have, if the works are to have their effect on us. This understanding need not be explicit, nor indeed could it become totally explicit or we would in effect be suggesting that the experience of such works is translatable into words. But I am suggesting that in understanding a work of music or architecture one sees it as expressive in a way that goes beyond its literally ascribable properties. One sees it as having a particular 'face' as I said in the previous chapter. And the way one justified one's understanding here is not a matter of pointing to objectively identifiable features of the work, so much as suggesting to another that it is possible to experience objectively identifiable features of the work in a particular way, as yearning, noble, elegant or whatever.[13] The success of a composer or architect is at the basic level dependent on his ability to convey to his audience the way in which he intends it to be experienced, and, in this, the criterion of success is analogous to that of the representational arts and of poetic metaphors.

What I have tried to suggest in this chapter is the way in which an artist does not aim at a literal representation of reality, but that his concern is rather with the evocation of feeling, experience and human attitude to the world, and I have said a little about what such evocation might consist in and how it might be achieved and assessed in various fields. I have also spoken of the centrality of the communication of an experience to all the arts, in contrast to the aims of scientific and literal objective discourse. So far, I have been taking artistic success to consist in success in communicating a particular vision or set of attitudes or experiences. But this is only the ground level of success in art. There is also the question of the worth or validity of the vision or attitude or experience. To this we will have to return.

CHAPTER 6

The Common Pursuit and Individual Creativity

> Someone said: 'The dead writers are remote from us because we *know* so much more than they did'. Precisely, and they are that which we know.
>
> (T. S. Eliot, 'Tradition and the Individual Talent', in *Selected Essays*, London, Faber and Faber, 1951, p. 16)

Eliot is, of course, wildly optimistic here if he is talking of explicit knowledge. Most of us explicitly know so little of the dead writers, and modern education is conspiring to ensure that even that little is concealed from our children. Moreover, even great sensitivity to our past can at times impress us more with the strangeness of our ancestors than with their familiarity to us. George Seferis, for whom the past of classical Greece was as living a presence as for anyone, was constantly remarking how little we really understand of the men whom we call the Greeks.

Lapiths and centaurs

Standing in front of the statues from the west pediment of the Temple of Zeus at Olympia, one can indeed feel unutterably distant from it all. The scene depicted is the savage attempt by drunken centaurs at the wedding feast of the Lapith king Peirithoos to carry off and rape the Lapith women and boys, and the resistance of the Lapiths. Above this scene of brutality, and taking no part in it, stands Apollo, the god, his hand outstretched, looking down on, and perhaps about to calm the orgiastic scene below. And we wonder, what could make this scene, depicted in a manner that emphasises at once the brutality of the events and the physical beauty of the participants, a suitable subject for a religious art? It was, after all, situated on Zeus's greatest temple in one of the most sacred sanctuaries in the classical world. How far can we understand a people for whom such a work was a supreme expression of their religion, a religion, remember, of bloody sacrifice, oracles and much garishness and peasant superstition? Can we really understand it as religious?

But then, looking further, we may notice that the whole composition has a rhythm going through it, like a series of waves which climax at the still centre, which is Apollo. There is a rhythm even in the Dionysiac turbulence of nature, and above the turbulence stands Apollinian civilization, the civilization of the Greeks themselves. Nietzsche could hardly have wanted a better representation than this to encapsulate the message of his *Birth of Tragedy*, even though he could not have known of it; while Thomas Mann leads us to think of the dialectic of Apollo and Dionysus as pervading European literature, art and music since *Tristan und Isolde*. Even though we cannot share in the classical-religious solution to the dialectic and perhaps hardly understand non-metaphorically just what *religious* conception the stones convey, we can still see the *problem* as ours. And we naturally see the *forms* of the sculptures as not strange at all, as most subsequent European sculpture has been based on such forms. European visitors who know nothing of Olympian religion, Apollo or centauromachies look at the pediment with a sense of recognition that would be quite lacking from an encounter with Aztec or Hindu sculpture, for the ideal of beauty exemplified in the figures of Apollo, Peirithoos, Deidameia, the Lapith women and boys, and even in the manly figures of the centaurs, is one that is part of European iconography. The paradox of the Olympian sculptures is that at the level of material form there is no problem for our aesthetic sense, but when we move to their deeper significance they become problematic. We can understand the Lapiths and centaurs as symbolising the triumph of civilization over barbarism in general terms. But it is terribly hard to make sense of the facts that a nation at that time and amid such general barbarism and despotism should have conceived such a thing in these precise terms and that it should have seen the god Apollo (to them, always a highly ambiguous and inscrutable figure, incidentally) in such human form, and that any of this should have been the essence of their religion.

The contemporary crisis in the arts

Seferis himself wrote in his preface to his Greek translation of Eliot:

> When myth was a common feeling the poet had at his disposal a living medium, a ready emotional atmosphere where he could move freely and approach his surrounding fellow men, where he could express himself.

and he compares what he wrote with Matthew Arnold's words in his essay 'The Choice of Subjects in Poetry':

> The terrible old mythic story on which the drama was founded stood, before he entered the theatre, traced in bare outlines upon the spectator's mind; it stood in his memory, as a group of statuary, faintly seen, at the end of a long and dark vista.[1]

The crucial question we will have to face later on in this chapter is how far meaningful art can continue to exist in the absence of a common set of myths or a shared religious background.

That there is a contemporary crisis in the arts, and has been for some time, is obvious. At the very least it should be worrying to all concerned that contemporary art has completely distanced itself from the taste of the generally informed and educated public. (To be more accurate, we should speak here of what is *called* contemporary art, for there is, of course plenty of contemporary 'non-contemporary' art, which professional critics and the administrators of artistic foundations and public museums and institutions of sponsorship choose to ignore.) Alongside this severance of publicly funded and lauded artistic endeavour from any basis in the appreciation of the only public that might be capable of sustaining healthy artistic endeavour, has been a wilful eschewal on the part of the very same critics and administrators – who currently constitute the metropolitan artistic establishment – of qualitative judgments on works of art from a broadly human perspective. They prefer instead to concentrate either on purely formal characteristics of works of art, or on making sociologico-political statements about their genesis and their relation to the 'society' which has produced them or to which they are a response.

Illustrating all that is worst in contemporary criticism a not always unperceptive art critic of a national newspaper stated in a review of a book by Sir Roy Shaw, a former Secretary-General of the Arts Council, that

> What Shaw misses here and everywhere is that [a] trite popular song performed with passion is every bit as good for you as ... Beethoven. *Indeed, they are different forms of the same musical experience.* (My italics.)[2]

The italicized sentence alone should cost the critic his job, for it demonstrates that he does not understand what artistic experience is: how the experience of an object (of any sort in fact, but including a work of art) is not like a sensation, such as a tingle or an orgasm, but is directed to the object which it is the experience of in such a way that it is the experience it is because of the properties the object has or is supposed by the experiencing subject to have. This last phrase opens the way for talk of more or less accurate, more or less informed, more or less rich, more or less humanly resonant experience of particular objects, while the manner in which different objects raise possibilities of different types of experience and different levels of meaning, complexity and inter-relatedness allows for qualitative discrimination between different objects, such as, in this case, the popular song and a Beethoven sonata. One might then want to suggest that the possibly greater immediate emotional appeal of the popular song is due to its psychological superficiality and musical simplicity compared to the Beethoven, which rewards study and repays re-experiencing in a way the popular song cannot; but this is the sort of judgment and discussion our critic has foreclosed with his belief that both give the listener the same experience. He is thus reduced to making purely technical-cum-formal remarks about works of art that essentially do not rise above the descriptive, or to locating their position in the forces of production of whatever society they have emanated from.

The critic is demonstrably confused about the nature of experience, but this does not explain why he has come to be so confused or why he should be predisposed to think Beethoven is as good for you as a trite popular song (his choice of words, remember). And this does take us back to what was said earlier about the inherited background of language, custom, evaluation and culture against which individual value judgements can be made, and to the questions we have been raising here about the role played by shared mythological and religious backgrounds in the maintenance of artistic traditions within which qualitative judgments of works of art can be made. It is, I think, not coincidental that the modernist crisis in the arts has coincided with the breakdown of traditional orders in other areas of society and the widespread repudiation in the West of religious faith.

Artistic modernism

Much of the impetus for the modern movement in the arts came from an assumption on the part of some artists that the traditions of Western art and culture which had reigned until the end of the nineteenth century had, for various reasons, been rendered obsolete. The reaction of the artists in question took the form of a search for purer forms of expression, which could cut through what was seen as the dross of the past clinging to the supposedly outmoded forms and styles prevalent at the turn of the century. Sometimes this search for a purer form of art emphasized expression and could be seen as an attempt to express one's feelings directly, eschewing all meretricious technique, (hence expression*ism*), and sometimes it emphasized the intellect, seeing the artist as the man who could reveal things in their true nature, uncluttered by ornament, decoration or by the merely contingent and historical. No doubt expressionism and the rigorous intellectualism of the later Schoenberg or of le Corbusier or of Mondrian look very different in the outcome, but they had a deep assumption in common, as is suggestively indicated by the fact that Schoenberg was not only musically an expressionist before embarking on his twelve-tone period, but actually saw his atonality as leading to purer forms of emotional *expression* than were possible under the old dispensation. What counted, he said, was the capacity to look deep inside oneself, freed from external order or convention. Deep inside one would discover one's true humanity. The assumption was that the twentieth century individual artist could free himself from the tradition in which he stood in order to achieve new forms of art from scratch.

Not surprisingly, ethical thinkers, scientists and philosophers of the period had similar assumptions in their own fields. The spirit of our age, at its start anyway, was one of despair at the heavy weight of the past combined with optimism about what could be done once one freed oneself from the past. Luis Buñuel well expresses the characteristically modern interweaving of revolution in the aesthetic with similar fresh starts in the moral and political realm when he wrote of surrealism, that most spiritually modern and humanly bankrupt of all artistic movements, that its founders conceived it as a 'revolutionary, poetic, and moral movement', aimed at 'exploding the social order' and transforming life itself. Not surprisingly, as Buñuel himself comments, the moral and political revolution sought by the surrealists never happened. And for this one must be grateful, given that as he says, surrealism was based on aggression and destructive impulses, more powerful than any creative urge.[3]

Against this quest for a pristine autonomy, in whatever field, and against the assumption of a pure pre-cultural human nature, we need to remember how much of what we do and think and are is the result of pre-existing language and custom and tradition, and that human beings are essentially social beings. A non-social, non-cultural human nature is a myth; and this is as true in artistic endeavour as it is in any other field. One may, of course, in a spirit of aggression and destruction repudiate much of what one explicitly knows of the past and attempt to throw off traditional constraints in one's work in favour of none at all, or impose on oneself new and untraditional ones. To some, such might seem an appropriate response to what one sees as a mendacious and inappropriate use of traditional forms by one's contemporaries. Ruskin mercilessly berated the citizens of Bradford for asking his advice about the style for the Exchange they were proposing to build:

> You cannot have good architecture merely by asking people's advice on occasion. All good architecture is the expression of national life and character; and it is produced by a prevalent and eager national taste, or desire for beauty.[4]

The fact they had to consult him as a 'respectable architectural man-milliner', to select for them which pattern from the past was most appropriate and fashionable showed that the conditions for a true architecture were lacking. And he loathed the use of Gothic in railway stations as being utterly out of tune both with the purpose of the railway and with the spirit of Gothic, adding that 'railroad architecture has, or would have, a dignity of its own if it were only left to its work. You would not put rings on the fingers of a smith at his anvil'.[5]

The most famous example of the allegedly shallow 'historicist' approach of the nineteenth century to architecture is the Ringstrasse in Vienna, where every grand style of European architecture but the contemporary is plundered for its monumental civic and imperial buildings; it is no exaggeration to say that the reaction against this paved the way for the modern movement in architecture and in the other arts.

While one can appreciate Ruskin's points and the drive for stylistic integrity which underlay the reaction to Ringstrasse architecture (which was not unmixed with distaste for what was seen as a corrupt and dishonest society and regime), we may now feel that neither the Bradford Exchange (Gothic revival) nor the Ringstrasse are as worthless aesthetically as some of their contemporaries did on moral and political grounds. Indeed, most people far prefer them to the architectural fruits of the modern movement. If integrity is what is at issue in repudiating nineteenth century historicism in art and architecture, one may question what the new men of the twentieth century who consciously turned their backs on the past thought of as integrity.

In a typical statement from the cultural milieu of pre-1914 Vienna, Oskar Kokoschka wrote

> Isolation compels every man, all alone like a savage, to invent his idea of society. And the knowledge that every doctrine of society must remain a utopia will also drive him into solitude. This solitude swallows us in its emptiness.[6]

If this is integrity, it is the integrity of Nietzsche's superman. For Adolf Loos, too, the function of art was 'to shake people out of their complacency (or comfortableness)' (and for this reason, converging strangely with Ruskin he opposed the application of art to the public domain of architecture), while Schoenberg declared that truthfulness to the inner dictates of mind and instinct, rather than beauty, were what mattered to the artist. And, a little earlier than the revolutionary works of Loos, Kokoschka and Schoenberg, Gustav Klimt had scandalized Viennese society by depicting 'Philosophy' on the ceiling of the new University hall in terms of a misty scene of anguished and passionate naked figures, rising from the depths in which we see the head of a priestess in a trance. As Carl Schorske points out, this is not the domain of a Wagnerian Erda, who is a warm, grief-laden, earth mother, but rather that of Zarathustra's Midnight Song, which affirms the dissolution of the boundaries of ego and world in the experience of cosmic suffering and desire.[7]

Expressionism has, of course, been a recurring theme in twentieth century art: the attempt to portray inner feeling unadorned and unsoftened by form or beauty. Earlier I quoted Adrian Stokes saying that art is sublimated desire and suggested that through art and culture more generally pure desire could be put at a distance and given different and more acceptable forms. But for many artists of the twentieth century, these forms have become unacceptable, with a consequent attempt to return inside to the desire itself in naked form. The paradox inherent in all attempts at pure expressionism is that, apart from unselfconscious cries and gestures, feeling cannot be expressed directly; artistic styles are one way of achieving indirect expression of feeling, while the values and concepts a culture and language make available to us in turn structure and form our actual feelings and desires just as much as any irretrievable basic instinct beneath the surface of our lives. The abandonment of style in art is less likely to lead to an authentic expression of one's actual feelings than to a self-dramatising display of adolescent brutality, in which one screams because, deprived of the stylistic means to express anything more choate, or articulated, one simply expresses that. Even where a great and sensitive artist like Mark Rothko transcends the adolescent gesturing and crudity of execution of a Schnabel or a Baselitz, and explores and draws us into a true inner space of real depth, what we are presented with in the end is an empty space, vibrant, it is true, but an emptiness nonetheless. He depicts feeling as mere possibility, awaiting enlivening form, in itself a passive nothingness: as bland and hopeless as the rectangles and plate glass of Mies van der Rohe's Seagram Building.

What, though, of the rationalism and intellectualism of much modern art, in the architecture of le Corbusier, say, or in the later music of Schoenberg? It is my contention that the roots, motivation and assumptions of such intellectualism are very similar to those of expressionism. In the first place, there is in both cases a rejection of ornament, beauty and traditional forms, in order to attain a supposed truth concealed by surface forms. Secondly, in both cases the contemporary artist attempts to derive a style from his own inner resources, feeling in the one case, intellect in the other. In both cases, there is a sense that the artist is a new, free man, standing outside history and society, and submitting his own individual taste and judgment for scrutiny, rather than simply forging a new link in a continuing tradition (as Cézanne thought of his work). For the expressionist, Kokoschka, the artist was a 'man without rules', breaking up the reality of his sitters' physiognomy in order to express the inner reality of instinct beneath; for the rationalist architect le Corbusier 'the Architect (his capital), by his arrangement of forms realises an order which is a pure creation of his spirit'.[8]

Le Corbusier propounded an architectural philosophy in which 'engineers' would dispense with custom and tradition, and even the appeal to utilitarian needs, in favour of geometrical forms, arrived at by calculation, which would satisfy 'our eyes by their geometry and our understanding by mathematics'. There was an analogous appeal to mathematical understanding underlying and even supplanting aural appreciation in Schoenberg's twelve-tone work, where many of the twelve-tone effects cannot be heard, but are perceived only by the eye or intellect in the study of the score. Very clearly, both le Corbusier, Schoenberg and many of their followers and contemporaries, including the constructivists in Soviet Russia, were influenced by quasi-scientific conceptions of art, in which what really counted was the form beneath and even concealed by the perceived surface, and which might only be valued by a new type of machine-oriented man in the future; this is the absolute antithesis of the spirit of the builders of the Parthenon, who famously departed from strict mathematical regularity in the construction and disposition of the columns in order that their building should *look* perfectly proportioned to the citizens of Periclean Athens.

In the case of the Parthenon, we have a good illustration of the way in which art works not on *a priori* or theoretical principles, but on perceived satisfactions of various sorts, with the emphasis on 'perceived'. Le Corbusier instructs us to 'eliminate from our hearts and minds all dead concepts in regard to the house, and look at the question from a critical and objective point of view (so as to) arrive at the "House Machine", the mass production house'.[9] The notion of the house machine, and of the man machine who will occupy it (a being depicted by Léger in terms of the rods, pistons and cylinders of the internal combustion engine and reduced to a scurrying stick insect in the architectural fantasies of Russian constructivism) epitomises the failure of much modern art and architecture, in its conscious and deliberate repudiation of the knowledge of perceived satisfaction in appearance and in life which has been encrusted in traditional styles, orders and moralities; these things have become traditional because they have, through experience and trial and error, been found amenable to human life. The so-called 'Purists' of post-1918 Paris saw themselves as traditional in that they constructed their works on geometrical patterns they discerned in the works of predecessors such as Poussin and Ingres. But in their obsession with mathematical purity and the mechanisation both of man and of the art-work, they succeeded only in their intention of eliminating from their work everything natural, comfortable, humane and practical for the way we actually are. The ready application of this type of style to totalitarian and bureaucratic purposes is no coincidence.

As to the 'discovery' of geometrical forms underlying masterpieces of the past, this is scientific reductionism of the worst type if it is used (as it was by le Corbusier and his purist associates) as a recipe for aesthetic productions, because it assumes that it is *because* of its geometrical form that a work of Poussin or Raphael, say, satisfies us. Once again, what is overlooked is the experience of the object. The experience of looking at a Raphael Madonna is precisely not like the experience of looking at a diagram of the golden section. To describe the experience of the former may involve some talk of harmony and balance, but the harmony and balance will be that of a human relationship, and of the relationship of the human to the divine. All this will be expressed through line and colour, but the significance of the lines and colours will be a human significance, expressive of the relationships between the woman, the child and the divine. It is, in any case, an empirical discovery that certain geometrical forms, rather than others have proved satisfying to the human eye in certain contexts, and this does nothing to show that even the golden section will always be satisfying in any context whatever, or that deviations from mathematical purity might not in other contexts strike the eye as equally well-proportioned. That this might or might not be so depends on the response evoked by the work itself and its context, and cannot be guaranteed by the application of mathematical theory.

Just as Nietzsche's revaluation of values was a failure and a philosophical absurdity – because the very values guiding the supposedly autonomous revaluer must themselves come from somewhere and remain unrevalued – so have the heroic attempts of modern artists to present themselves as visionaries, prophets and revolutionaries largely come to grief when in their works they have ignored the knowledge embodied in artistic traditions. Against the priest-like arrogance of a Walter Gropius, who revelled in the gulf between the public and the utopian artist, must be posed the true artistic humility of Wagner's Hans Sachs in *Die Meistersinger*, insisting that in the end, the public must decide. For in the end, the test of a work of art is the way it engages the untheoretical response of the public; a public with an education in feeling and taste, yes, but not necessarily in theoretical issues. Where there are theories underlying artistic forms and styles, as with the theory of music a music student learns, the usefulness of the theory must ultimately lie in its potential for producing works that are found perceptually, that is untheoretically, satisfying.

What I am claiming here is that the type of knowledge embodied in both art and morality is a practical knowledge, neither reducible to nor subordinate to theory. The example of the ornamental marbles is apposite here. The knowledge of the masons who group onyx, porphyry and marble together is not undermined by chemical theory, for the knowledge possessed by masons is a knowledge of the appearances stones can have in building. It is a knowledge of what is congenial and fitting to those who perceive and inhabit buildings and look at statues, not a knowledge of the chemical structure of things.

In a similar way, our knowledge of moral and political principles is not a knowledge of a pure spiritual order, fit for angels or *hommes-machines*. It is a knowledge handed down to us from the experience of previous generations of what is fitting for us, what makes life livable, what constraints are tolerable and liberties containable, what balances of order and freedom possible, what obligations enforceable without damage to individuality, what extent of duty and liability actually recognisable as binding on us, and so on. Much of this may appear irrational to theorists. Utilitarians, for example, find it hard to see why, if maximising pleasure and minimising pain are what we should aim at, special concern should be paid to maintaining and building up the integrity of one's family relationships, or close friendships, or why we should not, on occasion, torture the odd innocent individual to minimise the pain of many others. Those who make autonomous reason the criterion of morality have difficulty in acknowledging the extent to which our obligations, values and identities are given to us historically and contingently, by the families and communities into which we are born, the language we learn, the school we go to, the nation and time we are born into, and so on. It goes against the rationalist grain to admit that much of the fabric of our lives, including our espousal of specific values, depends on the particular circumstances and times of our lives, and that there is not always any absolute sense in which one way of structuring a society or set of values is superior to another.

If we see this point in philosophical terms, when we come to live, we would do well to respect the integrity of our own lives and their embedding in our actual values and society. As F. H. Bradley put it: 'the man into whose essence his community with others does not enter, who does not include relation to others in his very being, is a fiction.'[10] And similarly fictional is the rational and uprooted revaluer of all values, whether the values be moral or artistic. Uprooted and anomic one may be, but that in itself does not show that one's values are either uninherited or chosen in some completely free and rational spirit. The values of the Nietzschean superman are largely those of heroic society, which still formed a considerable part of bourgeois Christian morality in the nineteenth century, while I have already hinted at the Nietzschean legacy drawn on by some early twentieth century artistic revolutionaries reacting in their moral-cum-artistic crusades against nineteenth century comfort and smugness.

I am not, of course, saying that the *work* of all modernist artists (as opposed to some of their statements) is cut off from all past roots; in retrospect, continuities can sometimes appear beneath apparent revolutions and more striking than the breaks. Nor am I arguing that tradition, in either art or morality, has to be a static thing. I have already quoted Cézanne on this point. Many innovative artists in all fields have recognized both their debt to and their difference from their predecessors. And that is how it should be if a tradition is not to die. But we need to recognise the extent to which traditional knowledge in matters of art and morality is untheoretical and practical. Precisely because what is at issue in these cases are ways of perceiving and feeling, ways of reacting to the world and structuring one's perception and attitudes, the knowledge involved here is not like a matter of factual cognition, recognisable in purely objective uninvolved terms by people from any background. One's starting point here is always going to be the reactions, gestures, attitudes and types of perception one inherits, with all the historical contingency that attaches to such inheritances; that one sees and reacts to certain courses of action as shameful and others as incumbent on one, certain pictures or scenes as beautiful, others as desolate, certain uses of language as refined, others as vulgar, and so on. Such inheritances constitute the ways one finds one's way around the social world (as Bernard Williams puts it)[11] and the ways one perceives and reacts to a physical and social world, and form the raw material from which one creates works of art in response to it.

Because there is an ineliminable social aspect to both morality and art, both in learning one's criteria and in developing one's own responses and having them reacted to, it is hardly open to someone to strike out completely on his own. We talk about the modern *movement*, the *movement* in reaction to Ringstrasse architecture, and so on; it is only ideas which find resonance in the minds of others that properly condense in one's own mind, as Musil observed.[12] Because both morality and art involve, above all, a sense of what is fitting in the relevant context, actions and works of art will be judged by one's audience in terms of their perceived adequacy, rather than in terms of their meeting the demands of some preconceived theory. As the examples of the Parthenon and le Corbusier's *unité d'habitation* show, theoretical purity is neither necessary nor sufficient to command one's sense of the aesthetically fitting. (The one is aesthetically fitting and formally impure while the other is formally pure, and for many at least, unaesthetic.) More generally indeed, in art it is often precisely the departure from the rule which seems just right.

So far I have attempted to show that the modern movement in the arts has had both expressionist and rationalistic aspects. Both these responses to what has been seen as a failure of traditional styles and orders in the various arts have certain assumptions in common. In particular both assume that individual artists have within themselves the expressive or intellectual resources to start things anew, and indeed, that they have something like a duty to break free from the past in expressing their inner visions. The dilemma confronted by artists who reject the achieved order and tradition in an art form is to turn either to naked expressionism, in which the brutality of instinct and a neurotic absorption in one's overcharged psychology will tend to dominate, or to a new rationalism, simplifying human nature and the complexity of human relationships and deadening to human sensibility and, for these very reasons, highly congenial to political and commercial manipulators. Above all in a situation of breakdown of tradition, one will predict a multiplicity of 'movements' and manifestoes, and a plethora of personal 'styles'. But there will be little sense of a style as a continuing, organically developing force, linking the past to the present and making available to the present through the continuity in discontinuity of a new but traditionally rooted work, the knowledge gained in the past. Maybe today it is primarily in certain collective performing arts, such as music and classical ballet that this sense of a style which is not personal and not manipulable by individual whim or ingenuity or caprice still exists as a living force, and this because the public demands contemporary performances of the great works of the past (rather to the irritation of *bien-pensant* artistic administrators, it must be said).

The characteristically modern problem (or perceived problem) of the artist *condamné à être libre* and thrown on his own resources is not simply a result of hubris on the part of some modern artists. If they have often enough implicitly seen themselves as Nietzschean supermen,[13] revaluing values, this may be because they are responding to another aspect of the Nietzschean diagnosis: that according to which God is dead. What I mean here is that there is certainly no general consensus at present of a religious sort, nor any general agreement on the nature of man and his place in the universe. Indeed among many in our society there is a conviction that there could be no such thing, given that human beings are condemned to fashion their own world in the absence of any metaphysical or religious framework to provide a type of significance on which we could all agree. What I am suggesting is that we have to consider very seriously T. S. Eliot's analysis of the relationship between religion and culture.

Art and religion

In his *Notes Towards the Definition of Culture*, Eliot insists that he is treating religion and its relationship to the rest of human culture sociologically. In other words, he is not concerned there with questions relating to religion's truth. He is concerned simply to delineate the place religion has in informing a culture. Now, while it is clear that when there is a religion informing the life of a people or a nation, its impact will spread through the rest of that people's activities, it is not so clear whether Eliot thinks that there can be anything as cohesive as a culture or even a people without a religion. He writes 'any religion, while it lasts, and on its own level, gives an apparent meaning to life, provides the framework for a culture, and protects the mass of humanity from boredom and despair'.[14] He even says at one point that 'there is an aspect in which we can see a religion as the whole way of life of a people.'[15] Perhaps in line with this Durkheimian thought, he also speaks of contemporary Britain consummating 'its apostasy (from Christianity) by reforming itself according to the prescriptions of some inferior or materialistic *religion*' (my italics), and even of a new culture blossoming forth from this. [16]

In his essay, Eliot diagnoses particular forms of cultural disintegration. One case is when people cease to be organized in classes, which allow naturally for different levels of cultural achievement and aspiration. Another is when classes or other groups cease to communicate with each other, and high culture declines or ceases to enrich the whole of a society. Yet another is when the particular contribution people from specific regions can make to the whole community are stamped out in the name of some universalistic ideal. In general, the picture Eliot paints is of a healthy culture exemplifying unity in diversity and diversity in unity along several dimensions, social, cultural, geographical and even, up to a point, religious. One gets the impression from reading Eliot that he thinks that it is the religion of people which will provide the common bond of meaning which will make possible the appropriate balance of unity and diversity: and there is certainly something plausible in attributing to religion this type of role in a common culture.

Eliot insists that he does not want to identify religion and culture; nevertheless, he is emphatic that culture cannot be 'preserved, extended and developed in the absence of religion', and, crucially, that

> aesthetic sensibility must be extended into spiritual perception, and spiritual perception must be extended into aesthetic sensibility and disciplined taste before we are qualified to pass judgment upon decadence or diabolism or nihilism in art. To judge a work of art by artistic or religious standards, to judge a religion by religious or artistic standards should come in the end

> to the same thing; though, it is an end to which no individual can arrive.[17]

It is not just that a religion will provide a common stock of themes or symbols or myths for the artist, though it will do all of that, nor indeed simply that religion will set the framework for cultural endeavour, though it will do that too. (As Eliot points out, neither Voltaire nor Nietzsche is intelligible outside a Christian context.[18]) It is rather that religious standards are needed in order that art should have the authority many have claimed for it, to be a guide to life and taste, to be an informing presence in a civilization, rather than just a diversion for particular groups or cliques, fissiparous, decadent and ultimately trivial. When we look at the contemporary loss of confidence and absence of generally accepted standards exemplified by artists and critics alike, we might well feel that this is just what has happened to art today. It would also, in Eliot's view, be its religious ground and inspiration that allows us to think of Western art as constituting a tradition which we can recognize as binding and authoritative on us, that we are (even if only in the sense of Voltaire and Nietzsche) part of a religious tradition that has inspired and underpinned the works of art we see as constituting our great tradition of expression and self-understanding, and to which we may add our own works. For Eliot, it is not surprising that an irreligious intelligentsia looks to its art for the same sort of inspirational consolation that men of more faithful times have found in religion. For our artistic tradition is a tradition formed and informed by religion and Christian spirituality, and its spiritually authoritative and uplifting potential derives from its being in essence the expression of a religion.

If he did not see Western art as integrally bound up with Christianity, Eliot would have to agree with reductionist materialistic critics of Western art, such as Victor Burgin and Nicos Hadjinicolau. Such materialists attempt to demystify and de-aestheticize art, seeing works of art and their meaning as entirely conditioned by the forces of production in a society. For Hadjinicolau, without any religious sanction, all art could be is constituted by property relations and the assertion of the ideology of one class against another. The 'aesthetic' and spiritual values that bourgeois critics see in works of art are no more than a reflection of the attempt on their part to conceal through mystifications the underlying economic and ideological reality of the situation in which a work of art is made and perceived.[19] For Eliot, it is precisely from religion that we derive our fundamental sense both of the worth of things and of a sacred tradition encapsulating and transmitting that sense. It is from its connexion to religion that we are able to discern in works of art living expressions of an authoritatively based sense of the worth of things, and to place individual works of art within a tradition and judge them authoritatively against the standards achieved within that tradition. If there were no religious context to underpin the value we place on works of art, Eliot might well have, reluctantly no doubt, to concede to Hadjinicolau that aesthetic value is largely a matter of mystification.

I do not want to deny the truth of much of what Eliot says. We can feel at home in European art in a way we cannot with the arts of other cultures because of a shared European religious background. Even peoples as different from the Anglo-Saxon as the Russian are responding in their works of art to the particular problems raised for a nation and for individuals by adherence to the values of the Christian faith and, latterly, by the decline and replacement of those values by other values. Even as Russian a theme as that of Mussorgsky's opera *Kovanshchina*, describing the crushing of the Old Believers by Peter the Great, will be exotic to someone schooled within a Christian context only on the surface; for it is also a specifically Christian theme and deals with the impact of modern (or 'western') values on a traditional society and a traditional Christianity. The power of Mussorgsky's music and characterisation can almost seduce us even now in favour of the one and against the other, but in a way that we would hardly be seduced into a fundamentalistic Islam by a comparable Islamic work (if such a thing could be imagined). The reason for this is surely that the choice between Christian and 'enlightened' values is one we ourselves have been faced with (or if not we ourselves, our parents or grandparents), and one which we can understand and the resolution of which forms part of our actual self-understanding. Mussorgsky's genius is to convince us of the possibility of old Christianity as a way of life, and its values as values to live by, and to evoke the poignancy of its passing.

The challenge raised by Eliot to the non-religious is to show how art can retain its cultural and human significance and authority, and how an artistic tradition can be regarded as 'great' without implicitly borrowing authority from religion. For Eliot, the non-religious intellectual who looks to works of art for uplift, inspiration and a sense of the sacred is simply relying on the religious aura which still attaches to works of art for many in our society, but without any justification for doing so: he deserves the scorn heaped on him by the likes of Burgin and Hadjinicolau.

In my exposition and development of Eliot's position on the relationship between religion and art, I so far have concentrated on one half of Eliot's thesis: that aesthetic sensibility must be extended into spiritual perception. But Eliot also says that spiritual perception must be extended into aesthetic sensibility. If we pursue this thought we might be able to rescue art and artistic traditions both from the Scylla of Eliot's demands for a religious framework for art and the Charybdis of materialistic reduction of what, for want of a better phrase, we can call aesthetic values.

Aesthetic sensibility: 'mere' taste

What could be meant by aesthetic sensibility? At the very least, it must be a sensibility bound up with our embodiment and our modes of perceiving the world. It is sensibility which is sustained by our materiality. The urge to produce works for aesthetic contemplation is an urge to transform elements of the material world by means of human work. Even literature involves the use of language, which is itself a physical medium and is, as I suggested in chapter 5, firmly rooted in our embodied presence in a physical and social world. The very expression of inner states requires reference to our physical environment and outer form. From this point of view, the mystery is not to see how art and aesthetics could be materialistically vindicated, as to see how they could enter the realm of disembodied spirituality and universal value at all, when every moment of artistic and aesthetic activity is an assertion of our materiality, and every work of art does, as Marxist critics are right to insist, come from a particular society in a particular moment of history.

Further, as Peter Fuller has argued, art has played a central role in weaning Western man from attachment to a spiritual reality:

> The realism of Western oil-painting was ... certainly one of the ways in which men and women began to conceive of themselves

> in their own image, rather than God's ... In its very sensuality, oil painting helped to initiate an unprecedented form of imaginative, creative, yet thoroughly secular art which ... represents a genuine advance in the cultural structuring of feeling and expressive potentiality.[20]

Whether we regard this secular art as representing an advance or not, it was certainly part of the process by which western man began to explore and express his existence as material, embodied, earth-bound. The novel, by its very nature, centres on life on earth, personal and social, and it would not be wrong to see much Western music since the sixteenth century as secular in spirit, and appealing to our sensual nature (a realisation of which was behind the Catholic Church's repudiation at the turn of the twentieth century of harmonic music in favour of plainsong for liturgical purposes, as also, paradoxically, behind its espousal of harmony during the Counter-Reformation).

Our reconstruction of Eliot's position amounted to the assertion that an art which was not ultimately related to a religious perception of the world would lack the authority we see great art having over us. Aesthetic sensibility would, if severed from any roots in religion, become just a matter of taste, and the judgment that a Beethoven sonata was more valuable than a trite popular song, mere gesture, empty rhetoric.

One could, indeed, dispute that matters of taste are as arbitrary as this way of stating Eliot's position suggests. Take an example of literal taste. Someone who was unable to see that a fine Sauternes had qualities of taste, such as richness of texture and scent, which are absent in the average sweet white wine would be missing something that was really there; as well as speaking of the *pourriture noble* of the shrivelled grape used in Sauternes, one can employ comparisons with various other wines and the admittedly inadequate vocabulary of wine tasting to educate the uninitiated palate. There is, in fact, a convergence of educated taste on these points, as indeed there is on many things to do with food and drink.

A feeling that many people have on questions of literal taste in such areas as food and drink is that these things do not matter very much, and that there is something morally distasteful about concentration on them, particularly when one has pictures before one's mind of the effete wine snob on the one hand, and the starving

African child (or even the under-nourished Newcastle child) on the other. But this reaction is misplaced. Cultivation of genuine taste above all in one's day-to-day living is one of the primary ways by which people transform the merely instinctual and biologically determined, and can enter into genuine human practices, whose ends are not the simple satisfaction of animal needs. The convergence of taste on matters of food and drink, and the knowledge embodied in such things are pointers to something very important in human life: namely, the way we can and do imbue our animal nature and the satisfaction of our needs with imagination and taste. The fact that for some people this ability and the love of food and drink plays a disproportionate role in their lives does not show that it should not have its proper place in the lives of us all.

So we should not dismiss matters of taste in consumption as 'mere' taste or just arbitrary. The authority of good taste in food and drink is not arbitrary, but arises from the possibilities of response afforded to us by the interaction of our embodied nature and food and drink prepared with care, imagination and intelligence. And what is significant here for our purposes is that this authority is one firmly embedded in our materiality, which I presume no one would dream of referring to a spiritual source. It stands firmly on material feet.

It is natural, though, to think of works of art as pointing to something higher than our material nature, and natural, too, to think of them having a sanction in something more spiritual. Proust puts the point beautifully:

> There is no reason inherent in the conditions of life on this earth that can make us consider ourselves obliged to do good, to be fastidious, to be polite even, nor make the talented artist consider himself obliged to begin over again a score of times a piece of work the admiration aroused by which will matter little to his body devoured by worms, like the patch of yellow wall painted with so much knowledge and skill by an artist who must forever remain unknown and is barely identified under the name of Vermeer. All these obligations which have not their sanction in our present life seem to belong to a different world, founded upon kindness, scrupulosity, self-sacrifice, a world entirely different from this, which we leave in order to be born into this world, before perhaps returning to the other to live once again beneath the sway of those unknown laws which we have obeyed

> because we bore their precepts in our hearts, knowing not whose hand had traced them there – those laws to which every profound work of the intellect brings us nearer and which are invisible only – and still! – to fools.[21]

So in art, above all, we may hope to have revealed to us unknown laws, traced in our hearts before birth; this is what explains the intimations of depth, spirituality, heroism, understanding, care and knowledge we find in great art, and which is so often lacking from real life and often enough even from the real lives of the very artists to whom we turn for enlightenment.

One wonders, though, what talk of unknown laws and higher sanctions is doing in this sort of argument. Apologists on behalf of a rationalist architect such as le Corbusier are keen to point out that this building is based on the golden section, and that on the square, on laws, that is to say, of whose existence and influence most people are unaware in looking at the buildings. And this indeed, is as it should be. The buildings succeed architecturally and aesthetically in so far as they satisfy the people who live in them and who look at them. The lesson of modernist architecture is that reference to the invisible laws of golden sections, squares and the rest is so much Pythagorean mumbo-jumbo, because having some mathematically interesting structure is neither necessary for a building to succeed architectually and aesthetically, nor is it sufficient. This is not simply a criticism of modernist theorising about the arts. The same must be said of earlier attempts to base architecture or painting or music on mathematical formula, perhaps in the hope that in doing so one would, through one's work, commune with the mathematical essence of the world. In appreciation of works of art, it is not invisible laws which count, but appearances: how the building or novel or piece of music appears to those who are confronted with it. In this respect, Schoenberg and le Corbusier and their followers may be said to sin more than earlier artistic Pythagoreans, in the Renaissance, for example; for Renaissance attempts to mathematicize art did not entail the principled repudiation of delight in appearance, nor did it result in works that fail to delight.

In the way a work of art appears there can once again be degrees of success, in communicating what we might call truth, or, perhaps better, faithfulness to life, or to the experience of life. A vision which constricts, or fantasizes or over-simplifies or sentimentalizes or can only be enjoyed when drunk will lack the authority of one which respects the grain and complexity of life and experience, including, of course, its material basis. If a religious art of the past, and even more, of the present seems difficult to us today, this can be because in its striving after an immaterial perfection, it appears to evade our imperfect, embodied state and the problems our actual existence poses for us.

Practical knowledge and artistic tradition

Our view of what life is like, and how the world is perceived by us is given to us partly by biology and partly by history. This is true both of perception and of morality. Both no doubt have biological roots, and both are in various ways constrained by biology. It is impossible for us not to perceive the world in terms of colours, say, and not to be stimulated by certain sensations, tastes, smells, and so on. On the other hand, there is much in our particular modes of perception which is conditioned by our language, our society and, even by our art. In morality, one tends to stress social conditioning and the divergences between different moralities, many of which can plausibly be explained by historical circumstances. Nevertheless, any viable morality has to build on biological reality: on our biological needs and weaknesses, on our natural capacities for selfishness and altruism, on our natural store of sympathy, on our sense of kin and of kind. In so far as we have a sense of what life is like, and what faithfulness to it will consist in, much will be given to us by biology, on the one hand, and history, on the other. Any work of art which can satisfy us with being faithful to life must acknowledge the biology and history which makes us what we are. Neither the artist nor his audience starts afresh. Both intellect and feeling are rooted in the general facts of human biology and the particular facts of history which have made each what he is. Together, these are the factors yielding the very artistic styles and traditions to which we respond 'naturally' and which have played their part in forming our sensibility and sense of how things are. All this reflects itself in a practical untheoretical awareness and readiness to respond and to take the world and works of art in specific ways.

The re-valuer of values, be he moralist or artistic theorizer, is set on subverting this practical knowledge which we have as members of a particular species and as men in a particular moment of history by appeal to some rationalistic conception of his own. The twentieth century, post-Nietzschean modernist, exemplifies, as Gavin Stamp has suggested in writing of le Corbusier, 'all the destructive forces of the twentieth century . . . machine-worship, utopianism, collectivism, totalitarianism and megolamania'; personally, he is inspired by 'colossal vanity and egotism'.[22] These words are not, I think, unjust of one who sets his *a priori*, technology-dominated conception of how men ought to live and perceive their world up against what, as a result of biology and tradition, men have actually come to think and feel is fitting, is what life is and what it could be. In le Corbusier's own case, this refusal to bend to traditional knowledge or to compromise his mathematically inspired dreams reached extraordinary lengths, even late in his career when, according to the received view, he had softened somewhat his earlier commitment to building blocks of 'pitiless' magnificence (to use his own phrase). In the building of Chandigarh in the Eastern Punjab, not only did he fail in any way to accommodate the inevitable street traders of any Indian town (who came anyway), but he planned a rectilinear city of vast squares and freeways without any effective shade from the equally pitiless sun, and houses lacking the coolness provided by the traditional courtyard and terrace. Inhuman architecture was here allied to an equally utopian political vision, for Nehru had expressly ordered a 'new town unfettered by the traditions of the past' as a palliative for the loss of the beautiful, ancient city of Lahore.

The point one must make here against both the architect and the politician is that when we are attempting to respond to human needs and desires, the traditional always encompasses a vast pool of implicit knowledge, of procedures, customs, styles and designs which have survived because they have initially been found responsive to human needs and desires, and then constitutive of the needs and desires of succeeding generations. It is more than likely that until one disturbs a traditional order, one will not know just what the nature of the role of any particular element in it has been, because much of what is in any tradition will not have been planned and will not be there because it has been planned, but will have endured through a process resembling biological natural selection, shaped invisibly by its actual responsiveness to some need or desire or receptiveness that people have. The conception of tradition I am proposing is of an order emerging spontaneously through the survival of certain innovations and the destruction of others in terms of their actually answering or failing to answer the needs we actually have, rather than in terms of their initial apparent plausibility or self-evidence prior to testing. It is a notion which can be applied in many fields: scientific theorising, morality, law and custom, as well as in the realm of art.

But one lesson to be learned from the modernist movements in artistic fields in this century is surely the hazard of attempting simply to sweep away traditional forms and orders, by appeal to some rationalistic conception of what men ought to feel as opposed to building on what they have actually found and felt. In so doing, one sweeps away what knowledge *has* been gained about human perception and response. And the risk is compounded when one sweeps away old orders on the basis of *a priori* theories about human nature; for the knowledge embodied in an artistic tradition is essentially a practical knowledge, justified in terms of what people have responded to and found fitting rather than in terms of theory or its theoretical underpinnings. There may, of course, be theories *within* various artistic fields, such as the theory of harmony in music. But this type of theory and the use that is made of it is quite different from the way Schoenberg and le Corbusier conceived their theories. It is very much a codification of past knowledge and a device for initiating newcomers into a tradition, rather than something conceived of as a new basis for a purer type of expression, intellectually guaranteed. And unlike the theories of science – to which the theories of the modernists may be compared in their indifference to felt human response – a theory within an artistic tradition, whose role is to encapsulate the way a particular style has been found to work, will be tolerant of exceptions and allow development through invention and experiment. In contrast to the modernist use of theory in the arts, what has traditionally been spoken of as the theory of an art is actually the conservative, traditional aspect of that art. And a degree of conservatism in the arts, and a dread of utopianism is the only attitude which really respects both our biology and our history; it is no coincidence that among the epithets most commonly used against the work of twentieth century artistic revolutionaries are 'brutal', 'unnatural', 'mechanical' and 'utopian', for in their work they have often enough failed to respect either the nature of our embodiment and our place in nature or the actual historical and artistic context from which we start.

An artistic tradition is to be conceived in terms of its role in shaping our sense of what is fitting and natural. Such a conception underlines the need for contemporary artists to build on this sense rather than sweeping all the knowledge it embodies away in an attempt to create an artistic *année zéro*, based on pure theory and uncontaminated by our actual nature and history. But, in shifting the focus from the religious or mythological or metaphysical framework within which a tradition develops, have we side-stepped rather than answered Eliot's challenge as to the nature of the authority a 'merely' artistic tradition might have over us? But such doubts are, I think, unnecessary. A 'merely' artistic tradition, as I have been envisaging it, is a 'merely' human tradition, and one can read as much or as little into that 'merely' as one wishes. The point is that a 'merely' artistic tradition cannot be thought of or its values assessed in abstraction from what in general we think and feel about human life. The question Eliot is really raising is whether we can have any non-religious knowledge about human life and how it might be perceived and lived which might plausibly claim to be authoritative over us.

What I want to suggest is that there is, in any developed artistic tradition, a body of practical knowledge, which enables workers within that tradition to speak to an audience schooled in that tradition because and in so far as that audience can respond to its works as expressions of actual or possible experiences and attitudes, or as objects which will elicit from the audience felt and informed judgments about the value and adequacy of the works as objects of experience. Over and above the mere eliciting of an informed response, however, some works within a tradition will be authoritative over their audience in virtue of the quality of response they elicit: if they are perceived to be, say, beautiful or fitting or deep or intelligent or perceptive or tragic or richly comic.

If works of art are able to elicit such responses, and some account can be given of what it is about them which makes such a response appropriate in any particular case, I do not see why we need to look for any further religious sanction for their authority. Their authority consists in the fact that they are rightly perceived to be beautiful, fitting, intelligent, perceptive, tragic, and so on. And, as the example of much Greek and Christian religious art shows, works which were initially religious in inspiration can be found to be humanly authoritative even by those who do not believe in or even fully understand the underlying mythology. The Apollo at Olympia embodies a vision of human nobility and perfection of form we can still respond to, as does the characteristically Greek juxtaposition and rhythmical resolution of the beautiful and the violently tragic which we find in the rest of the pediment. And this, perhaps, is hardly surprising, for works such as this have, in virtue of their expressive power, helped to form our Western sensibility and sense of value, even though the substance of Olympian myth and religion has faded from our consciousness, and the ancient gods become as insubstantial for us as they were for Cavafy's Julian the apostate. In the case of Greek classical art, it is almost as if the material work and the human values and forms it embodies can survive the dissipation of the ideological mist which once encompassed it.

Immersed in an artistic tradition, what we learn is how to see and feel and how to express what we see and feel. We also learn those types of seeing and feeling which are appropriate in given situations, and this is where the aesthetic shades imperceptibly into the moral. We have seen in this century in many artistic fields a repudiation of traditional orders and styles in favour of naked expressionism or intellectual abstraction. One of the underlying motivations for modernism was undoubtedly a sense that old forms were incapable of conveying a new spirit or had sunk into spiritless formality or were being prostituted for unworthy ends. On occasion all or any of these criticisms might have been valid, but the remedies employed were neither called for nor conspicuously successful in communicating visions or works the averagely well-informed audience finds human or humane. And this was hardly surprising, given that it was just the knowledge of human expressive potential and of the nature of human life, encapsulated often unconsciously in the traditional styles, which had been repudiated.

The modernists, expressionists or rationalists indifferently, believed that in rejecting the expression of the past they could discover purer forms based on the inner truth which was revealed either to their feelings or to their intellect. But in this search for inner truth, they overlooked the fact that there can be no direct expression or even knowledge of inner feelings: we can express and understand what is inner only through publicly available forms. The characteristic crudity and brutality of pure expressionism arises from the fallacy of thinking that brutal and extreme feelings are the only true feelings. And here, of course, art itself becomes an uncivilizing force, contributing to a brutal and disordered climate. Modernists who seek the inner truth of art in pure intellect may be compared to the scientists we considered earlier who forget the embodied nature and consequent limitations of humanity. In the case of art, though, the forgetting of our embodiment is more destructive, for, as we have seen, in a traditional artistic style or order is encapsulated much practical knowledge of our embodied nature and its potential, which the rationalist would sweep away in order to construct his fantasies of new people and new worlds.

There have been, of course, great artists in all fields in the twentieth century who have been profoundly innovative in their work – and not 'mere' traditionalists – but who have not made it their business to indulge in the infantilities of expressionism or the arrogant intellectual bullying of the rationalists, and who are prepared to see themselves as continuing rich and inexhaustible traditions; in music there are Sibelius and Britten, Richard Strauss and Shostakovich; in the visual arts, Moore and Sutherland, and, in their own ways, Picasso and Matisse; in architecture, Lutyens; in dance, Balanchine and Ashton; and many figures, of course, in literature, of whom we need perhaps mention only Lawrence and Eliot. It has been customary in recent art histories to relegate twentieth-century 'traditionalists' to patronising footnotes and asides, so obsessed have we in the twentieth century become with the notion of futurity and the inevitable course of history. But such historicism (in Popper's sense of that term) is no more justified and no more admirable in art than in politics. If anything is right in the line of argument in this chapter, perhaps in future we should relegate the modernists and their 'post-modernist' successors (who see themselves as responding to the 'fact' that since the rise and fall of modernism the old traditions are well and truly dead) to the margins. Perhaps we should focus our accounts of twentieth century art on the endurance of traditional artistic styles and values through the collapses of morale and taste around them. But whether this is the way future histories will be written depends on creative artists now continuing to find meaning and sustenance in traditional forms such as tonal music, classical architecture, the realistic novel, figurative painting and classical ballet.

CHAPTER 7

A Ministry to Life

'Art is the best thing of all for veiling the terrors of the pit'.
Baudelaire

'Art saves him and through art – life'.
Nietzsche, *Birth of Tragedy*, Section 7

But how might art save? Nietzsche speaks in Section 7 of *The Birth of Tragedy*[1] of the 'metaphysical comfort' with which every 'true tragedy' leaves us 'that life is at the bottom of things, despite all the changes of appearances, indestructibly powerful and pleasurable': but do we need art to tell us this? Is this not rather the message of instinct, the joy of Zarathustra's Midnight Song, the dissolving, rather, of difference, of civilization, and of art? In the Kaufmann translation of *The Birth of Tragedy*, we find that Nietzsche quotes Wagner as saying that civilization is 'nullified' by music, just as lamplight is 'nullified' by daylight. This thought may be the message of *Tristan und Isolde*, but it is profoundly anti-art. Why do we need the appearance and the contrivance of art to effect a Dionysiac return to instinctual power and pleasure, in which art and every human bond and contrivance will be destroyed, as surely as in *The Bacchae* King Pentheus was torn to pieces by the Dionysian women on the slopes of Parnassus?

But did Wagner really say that, and is this Nietzsche's considered view even in *The Birth of Tragedy*? Wagner's thought is actually more subtle. The word Kaufmann translates as 'nullified' is 'aufgehoben', which has the sense not so much of nullification as of bringing to a fuller realisation through contrast. It is also the word used by Hegel to describe the way in which one ascends to a higher ground through first assertion and then denial. And Nietzsche goes on to speak of art as productive of saving illusions, turning nauseous thoughts at the horror of existence into notions with which one can live, and of art as taming the horrible and saving the will, just when it is in danger of being overcome by a sense that all is horrible and absurd. And this salvation, which doubtless involves an acceptance that one has biologically based needs and propensities, is effected by means of images of precision, lucidity and beauty sublimating the instinctual, as in the Olympia pediment. What we have is no mere subsidence into inchoate feelings, whether of joy or of suffering. Feeling can be transformed rather than destroyed through the devices of civilization and intellect, while intellect, contrivance and convention can be enlivened and their inherent constrictiveness, criticality and pessimism overcome through feeling. It is in art, par excellence, that the balancing act of the intellect that feels is most clearly manifested to us, and the two sides of man's nature harmonised. As Nietzsche says later, in *The Birth of Tragedy*, that 'truly serious task of art' is to deliver the subject from gazing into the horrors of the night and from the agitations of the will by 'the healing balm of illusion'.[2]

Self and world

It is possible to look at the dichotomies in human nature from a number of different points of view, each of which can throw some light on the nature of art itself, and the possibilities art opens up for us. From the psychological and psychoanalytical perspective, we can see the human individual as evolving into individuality from an original state of oneness with the mother. In this emergence into a separate individuality the child begins to form a conception of a world of objects, including its mother, as separate from itself, and threatening or benign as the case may be. But the original rupture and its painful effects remain, even where objects can be seen as benign. For some people a subsidence into an oceanic sense of oneness with the world is the only way to compensate for the pain of the original rupture. But this loss of identity, consciousness and sense of self, which is certainly gestured at in some forms of religious mysticism and in some works of art such as Rothko's colour field paintings, can itself be threatening and frightening. In a very real sense, one drowns in such a moment, as one is certainly sucked into and enveloped by a Rothko canvas. It is not surprising that the feeling Burke characterised as the sublime, in which one is enveloped by nature and which is closely akin to mystical transport, should be seen as evoking terror as well as pleasure. There is here an emptying out not only of self, but also of everything conceptualised and human. One gains what one has lost in becoming oneself, but at the same time loses what one has gained in becoming oneself.

While a return to an original organic unity with one's living environment would involve a loss of self and a loss of one's world, so from a different point of view does a fixation on the separateness from oneself of objects and the world. Such a fixation reaches its apogee in the version of the scientific enterprise sketched and criticised earlier. In Wittgenstein's 'pure realism' even one's own body and psychology form part of the world of objects, while the self retires beyond the boundaries of the empirical altogether, a mere observer of the world of objects. Here objects threaten the human by their indifference to us, and to our classifications and perceptions; and in a different but in just as real a way as in mysticism, the world is emptied of human meaning.

I want to insist that both the objectifying tendency and the unificatory urge are fundamental to our nature as human beings, and that some sort of balance between them is essential for our psychic health. We need to inhabit a world which is both structured by objects and suffused with our feelings. Such a world is what works of art are eminently suited to convey. Works of art can be, in the jargon of psychoanalysis, 'transitional objects' occupying the 'potential space' at the meeting point of the inner and outer: the point between which there is nothing *but me* (and hence, in a sense, nothing) and nothing but objects outside my control (and hence, no me). Of course, when a psychologist like D. W. Winnicott talks of transitional objects, he is talking of those things such as bottles, blankets and favourite toys which infants become attached to, and through which they effect their first halting negotiations between self and world – in a sense, constructing their selves and their worlds.[3] But although the mature person does not need a blanket, maturity does not resolve the tension between the self of pure subjectivity and feeling. We are always prone to slip back either into original formlessness or into the rigid, object-filled external world in which there is no conceptual space for subjectivity and selfhood and in which all is seen as subject to natural laws outside our control. Works of art are emphatically not blankets or toys, but by thinking of transitional objects and potential space we can come to see how a work of art can be a creative response to the tension I am just describing. In artistic endeavour the world of objects (or, in the case of music, *a* world of objects) is presented as informed by human subjectivity, and the feelings and attitudes that constitute this subjectivity are seen as structured as responses to the objects presented.

From the perspective of the building of a self properly integrated into an objective world, we can see why works of art should exemplify both objectivity and subjectivity. That is to say, what we need are works of art which are neither mere subjective fantasy nor eschewals of the subjective in the name of some supposed rationality or functionality. The world of objects is a world initially set against us and against our infantile demands. If art is to perform a redemptive role in our lives, it cannot do so by denying the existence of what psychoanalysts speak of as the bad breast which injures, cheats and frustrates us by being absent when we need it, or of our initially aggressive, frustrated and resentful response to the bad breast. It is indeed arguable that without the negative and alienating aspects of human existence, there would be no need for art at all, for in a work of art a human order is revealed in a world or in material which had hitherto seemed indifferent to human ordering. Both the resistance and the overcoming of the resistance are crucial in a work which is not mere fantasy, and this can fruitfully be linked to the basic fact of our embodiment in a world of objects, for it is this which provides both the ground for and the specific nature of our existence as material individuals in a world of material individuals.

The thought I am striving to express here about fantasy, resistance and objectivity in art has been well brought out in the context of sculpture and architecture by Adrian Stokes in his distinction between carving and modelling.[4] A carver is one who respects the integrity of his material, and who achieves his effects by articulating something which already exists in the block: 'The carved form should never, in any profound imaginative sense, be entirely freed from its matrix.'[5] Stokes speaks particularly of Agostino di Duccio's low reliefs in the Malatesta Temple at Rimini as examples of carving, as opposed to the modelling tendency prevalent in most nineteenth century academic sculpture, for example, in which something is 'fashioned' out of formlessness. Modern architecture is seen by Stokes as quintessentially a case of modelling; the 'inevitable' plasticity of new building materials has made it only too easy for modern architects to construct grandiose fantasies of their own devising, avoiding and repudiating the implicit knowledge of the traditional workers of traditional materials. In avoiding this knowledge, they have not only constructed works of subjective fantasy: they have also embarked on an art which, inspired by functionalist dogma, denies any space for human subjectivity at all, for the knowledge of traditional materials and traditional forms was first and foremost a knowledge of how things would actually look and feel to the human beings who were to look at and live in their buildings.

In practice, of course, the functionalism of modern architecture is not functional at all. Its works do not serve *actual* human needs and desires, and its functionalism is simply a demand that buildings *look* like geometrical figures or pieces of engineering. The modern movement in architecture is characterised by nothing so much as a repudiation of the aesthetic relevance of human subjectivity; and its world is as a result very far from being suffused by real human feeling and sensitivity. The example of modern architecture shows that there can be works which are at the same time subjective fantasy in their plastic approach to their material, and attempts to negate and deny our actual need for a humanly structured space responsive to our deep subjective needs – a perverse combination, indeed, and one that in practice comes perilously close to embracing a doctrine of art for art's sake. People will have the buildings we like the look of, whether they serve their needs or not. Only, talk of function has made it hard to see that the modernist movement in architecture has been first and foremost an aesthetic movement. (Le Corbusier's one redeeming virtue is that on the whole he did not attempt to hide this aspect of his work behind a mendacious vocabulary of functions.)[6]

By contrast to the brute concrete *unité d'habitation* striding peremptorily across some innocent and unsuspecting field, or the plate glass skyscraper blankly and over-bearingly ignorant of the stone city church it dwarfs, one may think of a classical Greek temple, that of Apollo Epikouros at Bassae, say. The harsh grey stone of the deep and remote mountains within which it stands is the very material from which its columns and walls have been fashioned. The temple seems to grow from the ground on which it stands, bringing focus and order into an otherwise untamed and, in Burke's sense, sublime landscape. It is as much a symbol of the juxtaposition of Apollonian and the Dionysian as the Olympia pediment a mere fifty miles away. At Bassae the god Apollo as saviour of humanity makes his presence felt in the heart of the wild, as one commentator puts it.[7] Here the otherness and impermeability of the natural and material have been respected in the working of column and entablature. Here the transformation and the ordering of the natural by the geometrical shaping of the stones and the layout of the temple express a deep harmony between the natural and the rational, which is forever foreign to the plasticity of modern architectural methods.

Thinking of the relationship between the ordered form of the classical Greek temple and the landscape, in some rocky fastness as at Bassae or at Delphi, or as at Sounion and Aphaia, where the temple is the crown of a mountain promontory overlooking the sea, a guide and symbol to shipping, or at Olympia where Zeus' temple nestles vast and splendid in the lush land between the confluence of two rivers, draws us naturally to thinking of another aspect of the duality in human nature, that between the natural and the conventional. Here once again, as already intimated, art can both express the tension and work creatively within it. The sense of the 'return to nature' implied in the classical orders, with their motifs based on natural forms, and even more in the Gothic, is something we need particularly to reflect on when science, technology and urbanisation are ever more rapidly estranging us from our roots in nature and our biological essence. Indeed, as Peter Fuller has observed, one of the most incongruous features of the simplifications of form characteristic of much modern architectural thought and also of movements in painting, such as futurism and constructivism, is that these simplifications are often justified by appeal to the supposedly simplifying tendency of modern science.[8] In fact, as we saw in chapter 3, it is only textbook physics which simplifies in the relevant respect, and that at a cost of accuracy and application to complex situations in the real world. In biology, by contrast to physics, far from simplifications of form or essence, what we find is a vast profusion of complexity, and also (some are surprised to learn, owing to the rather simplistic accounts of evolutionary theory that are current) of lack of functionality in organs and structure.[9] It is actually the lack of precise adaptation of species to present circumstances that may permit them to deal with new circumstances, so unfunctionality can be seen as having positive evolutionary merits. At all events, a proper scientific understanding of nature and our place in it would seem to argue for subtle and complex forms of art rather than for emulation of the gadgetry and geometrical forms of the machine age.

Art and the machine

Many artists of the twentieth century have, nevertheless, been obsessed by machinery and by analogies between human beings and societies and machinery (and the current form of this obsession is works of art based in techniques of computing and information technology). Andy Warhol said of his efforts to produce paintings that looked like prints that 'the reason I'm painting this way is because I want to be a machine'; he was speaking on his own behalf, but he could have been speaking just as much for many other twentieth century artists from the 1914–18 period onwards, who were equally obsessed with the image of its creator and his society projected by the machine. No doubt it is true that modern life has become increasingly impersonal and machine-like. The question is whether art and artists should simply acquiesce in the conceptions of man which result, or whether that task should be left to the advertisers and manipulators of taste and image (who are, in any case, better at it). Are we, in other words, to accept the art critic David Sylvester's fulsome tribute to Warhol's ultimate banality, his manifold of slightly different reproductions of the photographic image of Marilyn Monroe: 'In this art made out of what made its time what it was, the most significantly modern feature of all is that shifting of responsibility to the machine.'[10] I quote this judgment not just because of its characteristic historicising of art: *that* a work of art is worth notice because it is constitutive of its time; but also because in it lies the precise reason why, whatever else it might be, Warhol's *Marilyn* (along with countless other modern works) cannot be a work of *art*.

The reason I say this is because, according to Sylvester, 'responsibility' has been shifted to the machine. Our whole approach in this book has been to emphasize the way works of art are rooted in human subjectivity and human intentions, and can therefore be seen as expressive of the aims, attitudes and feelings of creatures who are able to form conceptions of their world and to shape their lives accordingly. To shift responsibility to the machine is to deny what we are and what we are capable of. Involved in such a denial may well be the replacement of forms of art with mechanical processes. The outcome of mechanical processes can be striking and, at times, even indistinguishable from works intentionally produced in every detail. But we must not make the mistake of thinking the two things are therefore the same, any more than van Meegeren could have produced a Vermeer, or, more to the point, that some randomly produced lines on paper could have the significance of the Rembrandt sketch they might by chance resemble. To do so would be to fall into the (characteristically modern) trap of conflating the result of an activity with the activity itself, and of thinking that the meaning of the one could be identified without remainder in terms of the meaning of the other.

A world given over to mechanical processes (such as ours largely is) may well contain a lot of images (as ours does, more in fact than at any time in the history of the world), but it will contain no art; no reflective, conscious manipulation of material to express and articulate a vision of the world rooted in human custom and sentiment. It might be urged that in so far as the mechanical image is itself guided and transformed in the processing stage or earlier by the imagination of its processor we do have a work of art, but as any honest photographer will admit, this 'in so far' should not blind us to the degree to which the detail and content of even the most tampered with and contrived photograph is not under the control of the photographer, in contrast to the way even the most mediocre artist is responsible for every detail of a painting or drawing. We value Rembrandt's drawings not just for their form which, as I say, one could envisage coming about almost randomly, but also for the psychological and visual penetration exemplified in presenting people and scenes so starkly in their visual essence.

The introduction into art of random and mechanical processes and, even worse, the replacement of art by such processes, such as we see in photography, computer produced 'poems' and aleatoric music, amounts to a repudiation of human beings as intentional agents, capable of creatively intervening in nature and guiding their actions in the light of their consciously monitored goals. As Rilke wrote *à propos* of the nineteenth century sculptor's pantograph:

> Long will machinery menace the whole of our treasure
> while it, unmindful of us, dares to a mind of its own . . .

Now it is life, no less, and feels best able to be it
having with equal resolve, ordered, constructed, destroyed.[11]

Rilke's words have proved prophetic. The introduction of mechanical processes into art is tantamount to a denial of our nature as creatures of flesh and blood, with feelings of flesh and blood (a denial which is quite explicit in the rod-and-piston creatures so beloved by the early Léger and his contemporaries). But in arguing that art should resist the seductions of mechanistic and electronic gadgetry, and the fantasies they provoke, and to respect the biological, including our own biological basis, I am not arguing in favour of a straightforward return to nature.

Nature

A savage Émile reared at the bosom of nature, whether noble or not, would be no artist, but a pitiable creature at the mercy of its instincts and those of its predators and enemies, actually locked in a struggle for bare survival. The fantasy of a sheer subsidence into the unformed, uncultured world of nature is a recurring artistic fantasy, delightful and perhaps necessary. One thinks of the scene in the *Magic Flute*, where on hearing Tamino's flute, the lion is at peace with the lamb, or of Siegfried's musings in the woods after killing Fafner or even of the Good Friday music in *Parisfal.* But such scenes are of primordial innocence or of redeemed existence, and are not of this world of nature. A natural world closer to our world is the infantile play of the Rhinemaidens liable to be shattered at any moment by the incursion of violent, malevolent and anarchic forces; such a world is tamed and made tolerable only by convention, law, contract – and art. And the role of art is perhaps crucial here, in its power to remind us of our roots in biology, while at the same time, transforming and re-presenting the natural in conventional form. A healthy art will show man of, but not submerged in, nature, at one with it, perhaps, but responding to it and reflecting imaginatively on it through culturally mediated forms. In this way one envisages Beethoven striding joyfully through the Viennese countryside, and expressing his joy in the supreme artifice of the Pastoral Symphony, which is an evocation of a properly human distance from and identification with the natural world.

Apollo and Dionysus

If art can help us creatively to express and explore the tensions within us caused by the fundamental dualisms of our nature, between the regressive self and the objective world, between feeling and reason, between the natural and conventional, between everything, in other words, which Nietzsche had gestured at in his initial distinction between the worlds of Dionysus and Apollo, it can also serve to break down the isolation an individual may feel from the feelings and sensibilities of the rest of mankind. It can do this by showing the individual that his own feelings are shared by others, and even more that his feelings are actually structured in the way they are by his belonging to a given culture and tradition. Through art, indeed, the individual can come to a powerful realisation of the truth of Bradley's claim that a community enters into his essence.

In *The Birth of Tragedy*, Nietzsche imagines tragic theatre originating in the activity of the chorus, in which individuality was initially submerged in a fusion of all present in communal song and dance. The whole throng begins to act as if transformed into Dionysus himself, with his primeval wisdom and knowledge of suffering. Eventually this knowledge is presented in dramatic, Apollinian form on the stage. Dionysian music is transformed into Appollinian images, and we recognize in tragedy a sweeping opposition of styles:

> the Appollinian appearances in which Dionysus objectifies himself are no longer "an eternal sea, changeful strife, a glowing life", like the music of the chorus, no longer those forces, merely felt and not condensed into images, in which the enraptured servant of Dionysus senses the nearness of the god; now the clarity of firmness of epic form addresses him from the scene; now Dionysus no longer speaks through forces but as an epic hero, almost in the language of Homer.[12]

This doubtless fanciful reconstruction of the origins of tragedy may find no direct analogue in the contemporary artistic scene, there being no civic form of art that corresponds to Greek tragedy. But we can surely agree with what underlies Nietzsche's myth: that a viable form of art has to touch on deep and universal human feelings mediated through images and forms accessible to a whole community. Only in this way can art really move us and give us the consolation and reassurance that we are indeed not alone, but members of a community of fellow sufferers and fellow revellers, and the necessary separateness of the individual and his perspective on the world brought about by our very embodiment elevated from the singularity of the 'I' to the commonality of a 'we'. Through art this transition can be made without losing one's individuality in an orgy of sensuality or destructiveness, precisely because in contemplating a work of art the observer makes his personal response to the work and is invited to test his subjectivity against that of the author of the work. In art, Dionysiac immersion is or ought to be tempered with Apollinian judgment.

So far in this chapter, I have been speaking of art in terms of certain dualisms which permeate our lives: those between a primeval, non-reflective self and a world of external objects, between feeling and objective rationality, between nature and culture. I have tried to suggest how in art both sides of these tensions are recognised, and elaborated and expressed creatively; and this is because artistic endeavour is based both in our fundamental embodiment and our ability imaginatively to transcend our immediate physical engulfment, and to reflect on it and other aspects of our lives. In art, we create works with an order and a narrative structure which would not be apparent in the even and relentless flow of physical time (or even in the a-temporal perspective of the absolute conception of reality). In art, we have the possibility of an imaginative response to our embodiment which, unlike science, does not seek to evade our embodied state, or the particular type of perspective we as humans have on the world. In artistic expression works can be created which reflect and speak to human beings as embodied consciousnesses, giving full weight to both sides of that duality.

In art, ideally there should be reflection of both our biological basis in existence and the transcendence of that basis which our reflective self-consciousness makes possible. My talk earlier, in chapter 4, of a 'metaphysical' self was intended to go no further than that: I have been treating our self-consciousness and our selfhood as an 'emergent' property of our evolution, taking us out of the purely natural environment, but not as capable of any existence without a material basis. It is in art above all that we get a practical expression of, and reconciliation of, both the material substructure of our existence and of our spiritual transcendence of that. The transcendence of the material implied in art is, like our existence itself, one constrained and circumscribed by certain material givens, such as the fact of birth and death, the nature of our perceptions, and our material needs and limitations.

Delightful illusions

Although I have so far referred to him only once, and have been writing in a very different idiom, both the general thesis of this book and the particular contrasts of the last paragraph are directly derived from Schiller's *Letters on the Aesthetic Education of Man*, which originally appeared in 1794–5.[13] For Schiller, the role of art is to find form in the formlessness of nature. Man does not fully exist until he transcends mere feeling, desire and impulse, and becomes a reflective self, affirming persistence within change, without at the same time annulling change altogether. We must develop both our capacity for feeling, 'intensifying passivity to the utmost',[14] and, at the same time, our capacity for reason and activity, so as to avoid an undue surrender to the formless and anarchic forces of sensuality. These two aims are realised simultaneously and harmoniously in the realm of art and of play.

In the Fifteenth Letter, Schiller expounds his central thesis that it is in play and in artistic creativity that the tensions between our sensuous and rational natures, our life drive and our form drive, as he puts it, can be creatively overcome. In play and artistic creativity we find a freely and humanly imposed order on material things, where the necessity and order is not extraneous to us, nor foreign to our embodied existence. The scientific and metaphysical tendencies of our mind, by contrast, foster greater independence of action than the artistic or aesthetic, but they do so by hardening our minds and depriving us of sensibility. In the Twenty Fifth Letter, Schiller speaks of truth as 'the pure product of abstracting from everything which is material and contingent' from the perspective of which we look on our feelings 'as something incidental, which could well be absent without the knowledge therefore ceasing to be knowledge or truth being any the less true'.[15]

On the other hand, our material existence reveals our limitations as the product of sheer external necessity, our desires and appetites as 'a monstrous round of ends' and the world as merely Fate. As with

Caliban, it is delight in beauty which first cajoles us from instinctive immersion in the material and our needs, and, the experience of beauty which continues to embed this reflective, self-conscious side of our being in the material world and our embodied nature, when we are tempted by science and metaphysics to soar above our embodied state altogether. The experience of beauty, in play, in art and in nature, is, in other words, the proof that freedom and harmony can exist alongside our dependence on and immersion in the material world.

In artistic endeavour we have not just the experience of beauty in the world. We also have a delight in the appearances (or semblance) we as human beings have imposed on matter. The semblance with which art deals in is not what Schiller calls 'logical' semblance, where an appearance is taken for the reality. It is a semblance which we know and recognise as human work:

> the reality of things is the work of things themselves; the semblance of things is the work of man; and a nature which delights in semblance is no longer taking pleasure in what it receives, but in what it does.[16]

But what it does here must not be confused with either the effort to describe reality exactly as it is or with the attempt to bring about specific effects in the world. To see art as either purely descriptive or instrumental would be to treat it as if it were not semblance, and this would betray its true nature, and compromise its essential freedom of spirit.

In his final letter, Schiller expands on his view of art as fundamentally non-utilitarian and based in delight in appearances for their own sake. He analyses the transition from uncivilised existence where pure need dominates to a more civilised form of existence in terms of an initial realisation of the way everyday objects can give disinterested and undirected pleasure through being given ornamentation. From the first realisation of the way in which things can satisfy an aspect of human nature which is not concerned with the satisfaction of basic need begins to grow a delight in civilised pleasures and forms, transforming first the outer and then the inner man:

> Unco-ordinated leaps of joy turn into dance, the unformed

> movements of the body into the graceful and harmonious language of gesture: the confused and indistinct cries of feeling become articulate, begin to obey the laws of rhythm, and to take on the contours of song.[17]

A harmonious, social realm grows up between the realms of constraint in physical and moral spheres:

> though it may be his needs which drive man into society, and reason which implants within him the principles of social behaviour, beauty alone can confer upon him a social character.[18]

Schiller argues that as taste for beauty is formed by a resolution of conflict between the two sides of man's nature, it brings harmony into the individual himself, and hence also brings harmony in society itself. Unlike the pleasures of sense which pertain to the individual *qua* individual, and so tend to separate men from each other, and unlike the demands of the moral law which (in Schiller's view) abstract from the individual as such, taste for beauty can unite individuals without removing them altogether from their sense of embodied individuality, and so can form the basis of a common feeling in society. This feeling will, in ideal circumstances, be genuinely common, open to all, from genius to child, from the noblest to the poorest born, transforming strength through grace, and curbing the defiance of the Lion by the bridle of a Cupid:

> In return, taste throws a veil of decorum over those physical desires which, in their naked form, affront the dignity of free beings; and, by a delightful illusion of freedom, conceals from us our degrading kinship with matter.[19]

Schiller admits that his vision of a republic of free and respectful individuals founded on taste and grace is fanciful, and that the redeeming power of aesthetic semblance is actually likely to be realized only in some 'few chosen circles, where conduct is governed, not by some soulless imitation of the manners and morals of others, but by the aesthetic nature we have made our own'. We would need to distinguish the type of chosen circle Schiller has in mind, which would be characterised by a respect for tradition and self-discipline from the shallow iconoclasm and hedonism of the Bloomsbury

Group, who also regarded themselves as a chosen circle, motivated by aesthetic values. We might also find Schiller's armchair anthropology fanciful, though perhaps not more so than those of Rousseau or Marx, say. For Schiller is surely right to stress the importance of aesthetic practices, or of aesthetic elements in practices, for a society transcending mere barbarism; a crucial element in human culture is the existence of practices embodying their own ends and standards, to which individuals must submit if they are to participate. And if in what he says Schiller underplays the extent to which these standards are historically given to us, rather than arising spontaneously from the intentions of individual novice participants, he is surely right to stress the way in which aesthetic taste and the perception of beauty are disinterested, rooted in our embodiment, and depend for their exercise on community between individuals in their experiences and reactions. On the other hand, a common feeling and a common taste is the source of a harmonious aesthetic community, rather than as Schiller suggests its effect. An aesthetic education is first and foremost an education in sensitivity towards existing standards and models.

Is Schiller correct, though, to see the cultivation of the aesthetic attitude in terms such as grace, dignity, simplicity, tranquility and nobility? It might be felt at the very least that there are certain subjects and certain epochs which would hardly promote such values or see works of art characterised by such qualities as an adequate or serious response to their time. Schiller himself was aware of this problem, but seems in a way to sidestep it:

> the infamous crimes of a Nero or a Commodus were put to shame by the noble style of the building whose frame lent them cover. Humanity has lost its dignity; but Art has rescued it and preserved it in significant stone. Truth lives on in the illusion of Art, and it is from this copy, or after image, that the original image will once again be restored.[20]

Schiller's idea is both moving and, in part, suggestive. A return to classical art and classical standards of taste might help us to restore lost dignity to our lives, or be a sign that we had already adopted a more dignified mode of existence than is currently popular. Although some independent minded architects are currently attempting to revive classical standards of taste in the face of hostility or indifference from schools and colleges of architecture, the task of improving the manners of a people through art is hardly attempted in the contemporary environment. And the danger of making such an attempt is that a superficial 'sweetness and light' will be seen to be mere preciousness, a pretty froth, in no way appropriate to the tragedy and suffering it for a time conceals. It might be argued, as Adorno argued, that our time is so riven with dissociations, contradictions and man-made horrors, that art itself has become problematic, a mendacious papering over what ought not to be papered over. The horrors of our time are such that there is something inauthentic and, in the last resort, frivolous even in the attempt to give expression to the horror in artistic form, or at least in any form which is not itself fractured or grotesque or utterly minimal.

'Our social situation'

Adorno's position is captured with unusual perspicuity in the following passage:

> In a work [of art], there is much seeming and sham, one could go further and say that as 'a work' it is seeming in and for itself. Its ambition is to make one believe that it is not made, but born, like Pallas Athene in full fig and embossed armour from Jupiter's head. But that is a delusion. Never did a work come like that. It is work: art-work for appearance's sake – and now the question is whether at the present stage of our consciousness, our knowledge, our sense of truth, this little game is still permissible, still intellectually possible, still to be taken seriously: whether the work as such, the construction, self-sufficing, harmonically complete in itself, still stands in any legitimate relation to the complete insecurity, problematic conditions and lack of harmony of our social situation; whether all seeming, even the most beautiful, even precisely the beautiful, has not today become a lie.

The passage is not by Adorno himself but is Adorno transformed and elevated by Thomas Mann into the persona of Adrian Leverkuhn – his Schoenberg-Nietzsche-Adorno figure – as mediated through the eyes of his pedantic narrator Dr Serenus Zeitblom.[21] Ironies and levels of meaning abound here, but as to the meat of the passage: I would agree that a work of art is seeming and sham, and not a mere copy of reality. And even showing the works and joints of a work (as in the 'happenings' of the 1960s) does not make a work of art any less a seeming or a sham. The question is, as to what sort of sham we want to be presented with in a work of art.

Claiming that the political blackness of our time precludes any but the grotesque or minimal in art and rules out any attempt at beauty or order in art not only issues in such absurd judgments on the part of contemporary critics as that a bronze cast of a heap of clay by the late Joseph Beuys is a 'triumphant creative achievement', the equivalent for our time of Rodin's *Gates of Hell*.[22] Like all attempts to interpret everything in historical or political terms, such a historicizing evaluation overlooks the way in which human activities have their own experiential and moral content, which is something over and above their political and historical context and irreducible to it, and which survives even in the bleakest and most oppressive of political circumstances. And in the case of art, the enforced reduction of what is allowable to a mere expression of its historical or political context is particularly unfortunate, because it is in art above all that our experience of the world can be imaginatively and morally enriched – and, in many cases, is so enriched by great works of the past. The fact that today we can respond to Homer and Greek sculpture, to Dante and Shakespeare, to Goethe and Beethoven is sufficient refutation of the reductionist thesis that sees all art as simply an expression of its own time,[23] for it shows that art can and does transcend contexts of history and politics. If one thinks further of an artist like the icon painter Andrei Rublev, who produced images of extraordinary purity, gentleness, spirituality, grace and strength at the end of an utterly desolate period of Russian history (the so-called Tatar yoke), we can see how art can transcend its time in every way as bleak as our own, and in many ways more so.

The aesthetic imagination is no passive reflection of the given world, but a primary mode in which we become aware of what we might become, by super-imposing our conceptions on the external world. For artists simply to acquiesce in the actual state of things, however oppressive this may be, is to repudiate that very power of imagination which is the root of all art and in a way to accept the imperialist claims of politicians on their own terms, instead of bearing witness to the way so much of life should be seen and conducted as independent of the dead hand of politics. Similarly, the attempt to use art as a handmaiden of actual political causes is equally misconceived. This is first because most of human problems and experiences and tragedy will continue just as they do now and have always done so, in different or even ideal political circumstances, and have very little to do with politics. We must, then, reject the claim often made by politicized aestheticians of the left that the usefulness of art may come to an end with the end of bourgeois society.[24] And secondly because in the best art we see that free flow of the imagination which would be entirely perverted were art and artists to have to labour for some predetermined secular cause, and to produce what would, in effect, be mere propaganda.

Decadence

At the same time, it must be admitted that while the human imagination can be used to enlighten and free us from the grosser aspects of our nature – the nakedness of our greed and need – it can also be used to oppress us with these aspects of our lives. Art can treat human beings and human life in terms of the animal in them, without allowing any scope for graciousness, magnanimity, friendship, love, dedication to something other than one's own animal desires, sobriety, or the other virtues. If anything is significant of our age and of the contemporary artistic scene it is surely that painters who shout loudest and paint biggest in glorifying the animal and the grotesque in human life are the current stars of the artistic establishment. I have in mind here such artists as Francis Bacon, who represents human beings as skewered carcasses and writhing meat,[25] Franco Clemente, Georg Baselitz, A. R. Penck and many of the other contemporary expressionists represented in the recent 'New Spirit of Painting' and 'German Art in the Twentieth Century' exhibitions, to say nothing of the appalling Gilbert and George with their cruel and narcissistic fantasies. Now, it may well be that the artists I have mentioned are not themselves conscious of this implication of their work and, despite their colossal commercial success, with such as the Saatchis, regard themselves as cultural critics. Some indeed explicitly see themselves responding to the post-Auschwitz, post-Hiroshima world by producing vast canvasses of unparalleled expressive crudity and ugliness. But to respond to evil with yet more ugliness is, in a way, to connive in it, to suggest that the evil is the whole story. The theory that the world is essentially evil redounds to the advantage of the world as it is, as Adorno himself said at one point.[26]

It is true that Nietzsche did want art to plumb the depths, but at the same time he saw the possibility of art presenting us with saving illusions. On the tyranny of modern art (by which he was referring to the work of Zola and Wagner), he wrote:

> Modern art as an art of tyrannizing. A coarse and strongly defined logic of delineation; motifs simplified to the point of formulas. The formula tyrannizes. Within the delineations a wild multiplicity, an overwhelming mass, before which the senses becomes confused; brutality in colour, materials, desires.[27]

This is certainly defamatory in respect of about 90 per cent of Wagner. Whether it is justified in the case of Zola I would not care to say, but it is a pretty accurate analysis of many of the German paintings the Royal Academy chose to represent post-war German art. Nietzsche may well be right in seeing ugliness unrelieved and a fascination with it as a symptom of decadence, a stimulus for weak nerves. The softer a society is, the more fascinated it becomes with the negative aspects of life. We see just this decadence in some of the recent ballets of Kenneth Macmillan from *Mayerling* onwards: decadence of this sort is the very antithesis to that divine gaiety, suppleness and wit, fire, grace, tenderness and even frivolity of the dancer which, following Schiller, was Nietzsche's own metaphor for the true artist and the true thinker,[28] which is so characteristic of the work of Macmillan's great predecessor Sir Frederick Ashton.

An art such as that of Schiele or Genet, which is obsessed with the unrelieved presentation of ugliness and squalor, is a moral outrage, because in it the artist puts his imagination at the service of a decadent and despairing attitude to existence. The more powerful the art is the worse it is morally. It may actually be some sort of existential contradiction of the imagination itself; the very faculty by which we can transform our embodied state into a semblance of something better is here being used to oppress us, to do dirt on life, as Leavis used to put it. I would, in other words, endorse what Anthony Savile says in trying to explain what it is which enables us to see great art in one depiction of the tragic or the fragmented and not in another. In the one there is merely a vicious spiral of aggression, where in the other, there is an ability to bring things together in healing unity.[29]

Redemption

While art may be able to bring together conflicting tendencies in human nature and to unify apparently opposing elements, including good and evil, it cannot do this by mere fantasies of sweetness and light, in which the negative aspects of life, or the hard realities of our embodiment of our condition as labouring, suffering beings are simply willed away in wish-fulfilment. It is this fault, above all, which condemns so much of the output of the appropriately named dream factories of the mass-media. Cinema and television can so easily be turned to the production of mere fantasy, precisely because the photography appears to put us in touch with a dense and completely detailed reality, while there is no such constraint on the content of the story thus projected. My thought here is well captured by Leavis's remark about Eliot, that he was a disturbing force and *therefore* capable of ministering to life.[30] Much of the greatest art helps us to understand that our sorrows and frustrations are, in Rilke's words

> nothing else
> than our winter foliage, our sombre evergreen, *one*
> of the seasons of our interior year – not only
> season – they're also place, settlement, camp, soil dwelling.[31]

As with the other dualities in our existence, we can, through art, come to some resolution of the tension, in seeing lamentation and jubilation as mutually dependent, in seeing how to find some order, even, in the disorder and waste of our lives. Art and aesthetic order and enjoyment can for a time veil the terrors of the pit. This they do not by veiling the pit through fantasy, but by balancing and ordering the terrors through art and artifice. Proust redeemed himself through art not by remembering the past, but by imposing order on it and by seeing an interweaving in life between pleasure, joy and fascination, and their obverse. As with a creative exploitation of the other dualisms in our lives, which arise from the basic fact of our natures as half animal and half angel, art is peculiarly well placed to counterpoise the morally and emotionally negative and positive aspects of our life. Judging a work of art from the perspective of life will involve at least some reference to its success in this task: as neither doing dirt on life, nor fleeing the real and the harsh altogether. In this way, art may play a significant role in forming and inspiring a civilization.

Saying that art can be an inspiring spirit in civilization takes us back to what we said in chapter 4 about the incompleteness of human nature and the way our horizons are not given to us purely by biology or closed off by our history. There is, for each of us, a space which is open, a space in which we can place our hopes, aspirations, visions and perceptions. Scientific theory, in its essential incompleteness and incorrigible idealising of the concrete cannot block out this space, except by a deliberate decision on our part that it should. If religion is not a living possibility for many today, it is left to art to inform the open, spiritual aspect of our life; not to deny biology or history or the material facticity of our lives, but to work through them and to show the spirit working through them, in embodying and expressing our hope for a more harmonious form of life, in which the various tensions and strains within our paradoxical form of existence are brought into some form of unity.

It is easy to see Rilke and Proust in this light, but I would suggest that in their different ways many great artists have brought harmony into the potential disorder and centrifugality of our existence, whether it be Sophocles finding a balance and a form of justice in the terrible atavistic story of Oedipus, or Rembrandt transforming the carcass of an ox into something solemn, grave and hieratic, with its gravity enhanced by the living flesh of the maid half-hidden behind it, or even perhaps Baudelaire's astonishing and dangerous performance in *Une Charogne*, in evoking the 'music' of a rotting corpse. Such transformations of the tragic and of the inevitable corruptibility of living strength and beauty are the highest manifestations of human freedom. The imaginative faculty, which is the basis of artistic activity can also be used in other, more negative ways: to oppress us with evil, as in Genet, or with meaninglessness as in Camus's *L'Étranger*, or to crush and intimidate the individual with inhuman size and form, as in ancient Assyrian sculpture. Artistic imagination can also be used to distract us from how things really are, as in much courtly art, or to trivialize and sentimentalize, as so often in popular fiction, or to rail hopelessly or cynically at the futility of existence and the works of man. In these and other ways art can deny our and its potential for harmony, and for the freedom which can be found in the creative reconciliation of opposites and in the ordering of the complexities of our lives. Such denials of human spiritual potential in art are as much revelations of the imaginative freedom we are endowed with as in the discovery of possibilities of space and harmony and balance in our lives. But it is only in attempting the latter, more difficult task that the aim of a work of art corresponds with the ground of its possibility.

Conclusion

We do not live in a world of our making. Our existence and survival depends on our recognizing this fact, and in coming to terms with the world as it actually is, not as we would like it to be. In natural science we have been able to develop a systematic account of many real regularities in the real world. Doing this has involved the correction of more primitive categories and classifications of natural phenomena which were implicit in our perceptual apparatus or explicit in our language, and also of earlier ideas of the nature of things. The pragmatic aspects of science are closely bound up with the correction of incorrect perceptions and theories. The successful prediction and manipulation of nature presupposes that there is at least some truth in our ideas about nature.

But scientific enquiry is not only a pragmatic matter. Submission to the world as it is involves a moral commitment to truth, and, in practice, the virtues of patience, humility and self-discipline.

Involved in scientific thinking are ideas which will always go beyond what is observable or checkable. It would be quite in order to follow Mary Hesse in regarding such aspects of scientific thinking as mythological or ideological. But we can do this without denigrating the success science has had in uncovering previously unknown regularities in nature and in grouping things together in natural kinds. We sometimes have good reason to think our classifications may actually have hit on groupings of phenomena which are based in the way things actually are, and not just in how they appear to us. Nor should we denigrate scientific knowledge or the quest for it, if we care about truth or the values involved in searching for and submitting ourselves to truth.

There is, nevertheless, a mythology about science which needs to be resisted. This is the idea that scientific knowledge is the only worthwhile type of knowledge open to us, and that the world is just what the currently accepted grand theories of science tell us it is. We have seen reasons to remain sceptical about the claims of grand theories of science where they go beyond natural regularities and classifications we can check, and no reason to suppose that they can radically undermine the validity of a picture of the world focused on human consciousness and its way of perceiving things.

We do live in a world that is not of our making, but in our lives and in our theorising we are constrained by our human perspective. We have also built up worlds of culture and of meaning, through which the implications of human perspectives on the world are explored and developed in various directions and within which we live out our lives as creatures of passion, flesh and blood.

It is important to realise the extent to which these worlds of culture and meaning are based on practical rather than on theoretical knowledge. Both the aesthetic and the ethical practices of a society will contain much that has been adopted without conscious decision or forethought, but rather because certain dispositions, values, arrangements and styles have evoked a favourable response in the society in question, and have complemented other elements in its life. One of the most insidious features of the scientific spirit is its tendency to suggest not simply that its knowledge has a unique standing and that anything that is really knowledge must stem from science, but that all true knowledge must conform to its ideal methodology of presenting its findings in terms of explicit theories and their deductive consequences.

But science itself is a human practice, one practice among others. Examination of its theories and methods does not support the view that scientific knowledge is specially privileged or free from unprovable and mythological elements.

Nor by virtue of their form and scope can the theories of science be used to express or develop our sense of value and of the way life is to be lived. On this aspect of the relationship between science and culture Leavis was quite correct. Art, on the other hand is intimately involved in our sense of the value of things. First, by means of its sympathetic re-enactments of anthropocentric perspectives on the world, it can play a central role both in value inquiry and in coming to an understanding of the nature of one's own existence and the meanings available in it. And, then, through its ability to resolve, at least for a time, certain fundamental tensions in our existence, it is well fitted to play a role in fostering harmony in one's own existence. Art can do this not so much by the advocacy of a quietistic acceptance of the way things are as by a creative and imaginative endeavour to discern harmony where once only discord and disorder were to be seen, and so to pursue the search for what David Bomberg used to call the spirit in the mass.

Notes

Introduction

1 F. Nietzsche. 'Attempt at Self Criticism' (Section 2), the new preface to the 1886 edition of his *The Birth of Tragedy* of 1872. (The words occur slightly differently on p. 19 of the Kaufmann translation of *The Birth of Tragedy*, New York, Vintage Books, 1967.) It is worth noting here that although the figure of Nietzsche plays an important role in this book, he stalks its pages as villain and hero in equal measure.
2 Vaclav Havel, *The Anatomy of a Reticence*, The Charter 77 Foundation, Stockholm, 1985, p. 13.

Chapter 1 The Scientific and the Artistic

1 Such anthropocentrism would, though, be pure grist to Heidegger's mill. For Heidegger the objects that make our statements true or false exist as such only in and through the disclosures conscious beings make of them, all otherwise being formlessness and shapelessness. We are so conditioned to seeing the world as *we* constitute it that we forget that what we see is just that: the world as we constitute it, a thought that will be familiar to readers of Donald Davidson's 'On The Very Idea of a Conceptual Scheme' (in his *Inquiries into Truth and Interpretation*, Oxford, Clarendon Press, 1984), though expressed there in a rather different idiom.
2 The example is due to Roger Scruton. See his paper 'The Significance of Common Culture', *Philosophy*, vol. 54, 1979, pp. 51–70, at p. 64.
3 Cf. A. Eddington, *The Nature of the Physical World*, Cambridge University Press, 1928, Introduction.
4 In 'Empiricism in the Philosophy of Science', in *Images of Science*, edited by P. M. Churchland and C. A. Hooker, University of Chicago Press, 1985, pp. 245–308, at p. 258.
5 This rather priestly attitude to what seem, on the face of it, contradictions and enigmas is in Polkinghorne's *The Quantum World*, Penguin Books, Harmondsworth, 1986, p. 82.

6 On this point, cf. Colin McGinn's *The Subjective View*, Oxford, Clarendon Press, 1983, pp. 131–4.
7 E.g. Thomas Nagel, in *The View from Nowhere*, Oxford University Press, 1986, p. 81.
8 L. Wittgenstein, *Notebooks 1914–1916*, edited by G. H. von Wright and G. E. M. Anscombe, Oxford, Basil Blackwell, 1961, p. 82.
9 Cf. L. Wittgenstein, *Remarks on the Foundations of Mathematics*, edited by G. H. von Wright, R. Rhees and G. E. M. Anscombe, Oxford, Basil Blackwell, 1956 and 1978, 1.112–16, 1.140–53.
10 G. Frege, *The Basic Laws of Arithmetic*, transl. M. Furth, Berkeley, 1964, 1.XVI.
11 Patricia Smith Churchland 'Replies to Comments' (on her *Neurophilosophy*, Cambridge Mass., Bradford Books, 1986) in *Inquiry*, vol. 29, 1986, pp. 241–72, at p. 260.
12 On this point, the arguments of Flint Schier in his *Deeper into Pictures*, Cambridge University Press, 1986, pp. 141–52 are decisive.
13 Adrian Stokes, *The Stones of Rimini*, New York, Schocken Books, 1969, p. 220.
14 F. R. Leavis, *Nor Shall My Sword*, London, Chatto & Windus, 1972, p. 98.
15 Cf. John Ruskin, *Modern Painters*, vol. V, part IX, ch. XI, sections 5–6.
16 Andrew Wilton, *The Life and Work of J. M. W. Turner*, London, Academy Editions, 1979, p. 220.
17 Leavis, *op. cit.*, p. 62.
18 Marcel Proust, *Remembrance of Things Past*, vol. XI (translated by C. K. Scott Moncrieff), London, Chatto & Windus, 1969, p. 320.

Chapter 2 A Decentred Universe

1 My discussion of Aristotle's science at this point draws on J. Losee's *A Historical Introduction to the Philosophy of Science*, Oxford University Press, 1972, pp. 8–15.
2 Cf. Aristotle, *De Caelo*, 286b10–287b22.
3 Cf. Losee, *op. cit.*, p. 52.
4 Cf. I. Newton, *Opticks*, 1704, Bk 1, Pt II.
5 Cf. J. Locke, *An Essay Concerning Human Understanding*, 1690, Bk II, ch. VIII, sections 9–10.
6 Cf. G. Berkeley, *A Treatise Concerning the Principles of Human Knowledge*, 1710, Section 18; D. Hume, *A Treatise of Human Nature*, 1739, Bk I, Part IV, Section IV.
7 W. Sellars in 'Philosophy and the Scientific Image of Man' in his *Science, Perception and Reality*, London, Routledge & Kegan Paul, 1963.
8 Berkeley, *Principles*, Sections 10–15.
9 In *Central Questions of Philosophy*, London, Penguin 1976, p. 87.
10 Cf. Locke, *Essay* Bk IV, Ch. II, Section 11.

11 Cf. P. F. Strawson, 'Perception and its Objects', in *Perception and Identity*, edited by G. F. Macdonald, Macmillan, London, 1979, pp. 41–60, at p. 59. My discussion of the relationship between the scientific world view and the standpoint of common sense has drawn at a number of points both on Strawson's paper and on A. J. Ayer's comments on Strawson in the same volume, pp. 277–98.
12 Cf. A. Eddington, *The Nature of the Physical World*, Cambridge, 1928, Introduction.
13 The example is from Locke, *Essay*, Bk II, ch. XXIII, Section 11; the refutation will be found in Hume, *loc. cit.*: 'tis impossible to conceive extension, but as composed of parts, endowed with colour or solidity.'
14 In *Revolutions and Reconstructions in the Philosophy of Science*, Brighton, Harvester Press, 1980, p. 105.
15 *The Character of Physical Law*, London, British Broadcasting Corporation, 1965, p. 56.
16 Cf. Locke, *Essay*, Bk IV, ch. III, Section 13.
17 R. Dawkins, *The Blind Watchmaker*, Longman, Harlow, 1986, pp. 33–7.
18 T. Nagel, 'What is it like to be a bat?', *Philosophical Review*, vol. *83*, 1974, pp. 435–50.

Chapter 3 Science and Truth

1 *The Gay Science*, 1887, Section 300, translated by William Kaufmann, New York, Vintage Books, 1974.
2 Cf. K. R. Popper, *Conjectures and Refutations*, London, Routledge & Kegan Paul, 1963, pp. 97–101.
3 Cf. J. Losee, *A Historical Introduction to the Philosophy of Science*, Oxford University Press, 1972, pp. 83–4.
4 Cf. B. van Fraassen, *The Scientific Image*, Oxford University Press, 1980, p. 60.
5 P. Duhem, *The Aim and Structure of Physical Theory*, New York, Atheneum, 1962, p. 7. (Duhem himself opposed this view of scientific theory.)
6 Popper in 'Replies to my Critics' in P. Schilpp (ed.), *The Philosophy of Karl Popper*, La Salle, Illinois, Open Court, 1974, pp. 959–1197, at pp. 1192–3.
7 Cf. H. Putnam, *Meaning and the Moral Sciences*, London, Routledge & Kegan Paul, 1978, pp. 130–3. Real-life examples of debates between (for a time at least) empirically equivalent but theoretically competing theories are given by Larry Laudan in his *Progress and Its Problems*, London, Routledge & Kegan Paul, 1977, pp. 47–8. They are those between Copernican and Ptolemaic astronomy (1540–1600), between Newtonians and Cartesians (1720–50), between wave and particle optics (1810–50) and between atomists and anti-atomists (1815–80).
8 T. S. Kuhn, *The Structure of Scientific Revolutions*, University of Chicago Press, 1962.

9 In *Science and Scepticism*, London, Hutchinson, 1984, p. 202.
10 Kuhn, *op. cit.*, pp. 117–18.
11 Mary Hesse, 'Science Beyond Realism and Relativism', forthcoming in the Proceedings of the Conference on Cognitive Relativism and Social Science, Utrecht, 1986, manuscript, p. 5.
12 Cf. Putnam, *op. cit.*, p. 126.
13 Hesse, *op. cit.*, p. 6.
14 Cf. van Fraassen, *op. cit.*, pp. 19–40.
15 *Ibid.*, p. 40.
16 That philosophy of science has been, to its detriment, 'theory-led', while neglecting experiment, is one of the main themes of Hacking's *Representing and Intervening*, Cambridge University Press, 1983.
17 Mary Hesse, *Revolutions and Reconstructions in the Philosophy of Science*, Brighton, Harvester Press, 1980, p. xix.
18 Cf. K. R. Popper, *Objective Knowledge*, Oxford University Press, 1972, p. 212.
19 C. S. Peirce, *Collected Papers*, 1935, vol. 6, p. 35.
20 Nancy Cartwright, *How the Laws of Physics Lie*, Clarendon Press, Oxford, 1983.
21 *Ibid.*, p. 154.
22 *Ibid.*, p. 162.
23 K. R. Popper, *The Logic of Scientific Discovery*, London, Hutchinson, 1959, p. 277.
24 Cf. Pierre de Laplace, *A Philosophical Essay on Probabilities*, 1819, Ch. 2.
25 On the difficulties and paradoxes inherent in self-prediction, even within a deterministic, Newtonian universe, cf. K. R. Popper, 'Indeterminism in Quantum Physics and in Classical Physics', *British Journal for the Philosophy of Science*, vol, I, 1950, pp. 117–33, 173–95.
26 Bernard Williams, *Ethics and the Limits of Philosophy*, London, Fontana, 1985, pp. 139–40.
27 Bertrand Russell, *A.B.C. of Relativity*, Third Edition, London, Allen & Unwin, 1959, p. 34.
28 James A. Coleman, *Relativity for the Layman*, Harmondsworth, Penguin Books, 1959, p. 73.
29 This way of putting the matter is due to Thomas Nagel, in *The View from Nowhere*, Oxford University Press, 1986, p. 76.
30 Cf. P. T. Geach, *God and the Soul*, London, Routledge & Kegan Paul, 1969, p. 92.
31 Heisenberg and Bohr as quoted in K. R. Popper, *Quantum Theory and the Schism in Physics*, London, Hutchinson, 1982, pp. 40–1. Popper's book is a sustained and partially successful onslaught on the Copenhagen interpretation of quantum physics.
32 Mary Hesse, 'Texts without Types and Lumps without Laws', *New Literary History*, vol. XVII, 1985–6, pp. 31–48, p. 40.
33 Mary Hesse, 'Science Beyond Realism and Relativism', p. 9.
34 Marjorie Grene, *The Knower and the Known*, London, Faber & Faber, 1966, p. 33.

Chapter 4 Human Culture and the Role of Artistic Expression

1 Cf. *Human, All Too Human*, Section 519.
2 F. A. von Hayek, *Law, Legislation and Liberty*, vol. III, Routledge & Kegan Paul, London, 1979, Epilogue.
3 Cf. A. MacIntyre, *After Virtue*, Duckworth, London, 1981, ch. 14, esp. pp. 175 ff.
4 M. Oakeshott, 'Education: The engagement and its frustration' in R. F. Dearden, P. H. Hirst & R. S. Peters (eds.), *Education and the Development of Reason*, Routledge & Kegan Paul, London, 1972, p. 19.
5 Oakeshott, *op. cit.*, p. 21.
6 Cf. D. Wiggins, *Sameness and Substance*, Basil Blackwell, Oxford, ch. 6, especially sections 7–11.
7 Cf. F. Schiller, *On the Aesthetic Education of Man*, Clarendon Press, Oxford, 1982, p. 213.
8 Cf. Oakeshott, *op. cit.*, pp. 20–1.
9 *The Birth of Tragedy*, *loc. cit.*, p. 78.
10 On Proust and redemption through memory, cf. P. Munz 'The Evocation of the Senses by the Prose of Freud and Proust', to appear in H. Delbrück (ed.), *Sinnlichkeit in Bild und Klang*, Hans-Dieter, Heinz Akademischer Verlag, Stuttgart, 1987, pp. 415–30.
The standard biography of Proust (in any language) is George Painter's *Marcel Proust*, Chatto & Windus, London, 2 vols, 1959 and 1965.
11 F. R. Leavis, *Nor Shall My Sword*, Chatto & Windus, London, 1972, p. 207.
12 Leavis, *op. cit.*, p. 97.
13 Cf. Oliver Sacks, *The Man Who Mistook His Wife for a Hat*, Picador, London, chapters 22 and 23, esp. p. 198.
14 Scruton's views on music are elaborated in his 'Understanding Music', in his *The Aesthetic Understanding*, London, Methuen, 1983, pp. 77–100.
15 Deryck Cooke, *Mahler 1860–1911*, The British Broadcasting Corporation, London, 1960, p. 43.
16 Williams' views on the humanities are elaborated in his Raymond Priestly lecture 1987, a shortened version of which appeared in the *Times Educational Supplement*, 23 January 1987, p. 4.
17 The *locus classicus* of the expression of these difficulties with Marx and Freud is K. R. Popper's *Conjectures and Refutations*, Routledge & Kegan Paul, London, 1963, ch. I. I probably should not say that Marxism and Freudianism are just moral visions. Part of their charm is that they purport to explain why aspects of the human world are as they are, in terms accessible only to those who understand the respective 'sciences'. They are, of course, attempts to de-mystify parts of the human world by showing that the appearances there are as illusory as to their true nature as Newton held the common-sense view of secondary qualities to be. My point, though, is partly that the scientific status of Marxism and Freudianism is widely disputed on grounds of their lack of empirical success, and that this is why many have come to see them

primarily as moral visions. But even if they were taken as scientific theories, though, my line of criticism in the text would be pertinent, in raising the question as to whether they adequately *describe* what they then purport to explain away.

Chapter 5 Truth and Art

1 Cf. P. T. Geach, *God and the Soul*, London, Routledge & Kegan Paul, 1969, p. 72.

2 From 'Truth and Falsity in an Ultra Moral Sense', in *The Philosophy of Nietzsche*, edited by G. Clive, New York, Mentor Books, 1965, p. 508.

3 My argument on the primacy of the literal derives from David E. Cooper's *Metaphor*, Oxford, Basil Blackwell, 1986, pp. 275–9.

4 Cf. L. Wittgenstein, *Lectures and Conversations on Aesthetics, Psychology and Religious Belief*, University of California Press, 1972, p. 72.

5 *Ibid.*, p. 63.

6 In my *Experience, Explanation and Faith*, London, Routledge & Kegan Paul, 1984.

7 W. Otto, *The Homeric Gods*, London, Thames & Hudson, 1979, p. 232.

8 The use of metaphorical language to capture one's inner states has been well analysed by Peter Munz in his *Relationship and Solitude*, London, Eyre & Spottiswoode, 1964, pp. 58–9.

9 Cf. F. A. von Hayek, *The Sensory Order*, London, Routledge & Kegan Paul, 1987, p. 21.

10 *Iliad*, Book XI, lines 172–8, translated by R. Lattimore, University of Chicago Press, 1951.

11 John Haldane, 'Aesthetic Naturalism and the Decline of Architecture', in *Issue* 5, July 1986, pp. 1–13, at p. 5.

12 My analysis of musical experience is based on Scruton's 'Understanding Music' in his *The Aesthetic Understanding*, London, Methuen, 1983, pp. 77–100. The quotation on rhythm is from p. 90. On hearing some tones as high and others as low, Scruton points out that someone who did not hear tones in the way we do, and called our 'high' tones low and vice versa – as it has been contended the ancient Greeks and Chinese did – would fail to comprehend what are for us such elementary musical experiences as the opening of *Rheingold* or the soaring solo violin in the Benedictus of the *Missa Solemnis*.

13 Cf. R. Scruton, *Art and Imagination*, London, Routledge & Kegan Paul, 1982, p. 54: 'In aesthetics, you have to see for yourself, precisely because what you "see" is not a property .... To agree in the judgment that the music is sad is not to agree in a belief, but in something more like a response or an experience; in a mental state that is – unlike belief – logically tied to the immediate circumstances of its arousal.'

Chapter 6 The Common Pursuit and Individual Creativity

1 Both passages are quoted by Seferis himself in *A Poet's Journal: Days of 1945–51*, translated by A. Anagnostopoulos, Cambridge, Mass., Belknap Press, 1974, p. 45.
2 Waldemar Januszczak, in the *Guardian*, 30 January 1987.
3 Cf. Luis Buñuel, *My Last Breath*, translated by Abigail Israel, London, Jonathan Cape, 1984, pp. 107–9. On the tantalising connections between this aggressive nihilism and the espousal of Marxism, which was, of course, the creed of many of the surrealists, see Roger Scruton's chapter on Sartre in his *Thinkers of the New Left*, Harlow, Longmans, 1986, pp. 176–92.
4 John Ruskin, *The Crown of Wild Olive*, Lecture II, Section 54.
5 Ruskin, *The Seven Lamps of Architecture*, ch. IV, Section 23.
6 This and the references to Loos and Schoenberg which follow are taken from ch. 7 of Carl Schorske's *Fin-de-Siècle Vienna*, Cambridge University Press, 1981. While I disagree profoundly with Schorske's conclusions, his study is the best I know in laying bare the roots and aspirations of early modernism in the arts.
7 Cf. Schorske, *op. cit.*, pp. 239–40.
8 Le Corbusier, *Towards a New Architecture*, London, Architectural Press, 1978, p. 9.
9 Le Corbusier, *op. cit.*, p. 13.
10 F. H. Bradley, *Ethical Studies*, Oxford University Press, 1927, p. 168.
11 In *Ethics and the Limits of Philosophy*, London, Fontana Press, 1985, p. 168.
12 In *The Man Without Qualities*, London, Picador, 1979, vol. I, p. 134.
13 And, as in the case of the artists Joseph Beuys and Lupertz, explicitly too.
14 T. S. Eliot, *Notes Towards the Definition of Culture*, London, Faber & Faber, 1962, p. 34.
15 Eliot, *op. cit.*, p. 31.
16 Eliot, *op. cit.*, p. 34.
17 Eliot, *op. cit.*, p. 30.
18 Cf. Eliot, *op. cit.*, p. 122.
19 Nicos Hadjinicolau's views are expounded in his *Art History and Class Struggle*, London, Pluto Press, 1978.
20 Peter Fuller, *Seeing Berger: A Revaluation*, London, Writers and Readers, 1981, pp. 15–16.
21 Marcel Proust, *Remembrance of Things Past*, Scott Moncrieff translation, vol. IX, pp. 250–1.
22 Gavin Stamp, 'The Consequences of le Corbusier', *Daily Telegraph*, 9 March 1987.

something that must be striven for constantly. The more explicitly political expression of the 'end of art' thesis is found in Adorno:

> When classless society promises the end of art, because it overcomes the tension between reality and the possible, it promises at the same time the beginning of art, the useless, whose intuition tends towards the reconciliation with nature, because it no longer stands in the service of the exploiter's use.
>
> (Adorno, Theses on Need, *Gesammelte Schriften*, VIII, p. 396, quoted in Martin Jay's *Adorno*, London, Fontana, 1984, p. 100)

Precisely: if there were no evil and no conflict, there would be no art, or no useful art, as I argued in considering the problem of evil in my *Experience, Explanation and Faith*, (London, Routledge & Kegan Paul, 1984, ch. 5); but the superficiality of Adorno's position is to think that the tension between reality and the possible – or, I would say, the impossible – has anything essentially to do with bourgeois society.

Joseph Beuys (quoted in *German Art in the Twentieth Century*, p. 16) in like spirit to Adorno seems to have envisaged the present function of art as one of dismantling 'the repressive effects of a senile social system' prior to a time when we would build 'a social organism as a work of art'. He also speaks (*op. cit.*, p. 129) of 'the social sculpture which man as an artist is helping to build'. (Do we need to remind ourselves yet again that both Hitler and Trotsky saw politics in terms of art, as did Mussolini, who saw himself as a sculptor whose mission was plastic humanity?)

And we find the following Adorno-like, but even more gnomic utterance in Georg Bussmann's contribution to the *German Art in the Twentieth Century* catalogue (p. 123). Writing of those artists included in the Nazi's 'Degenerate Art' exhibitions, he says:

> What art has to offer is the initiation of reflection or a dialogue about history and the present and about man's place in them. It is ultimately the chance of freedom, freedom which resides in our perceiving the whole cultural spectrum as an arena in which we can move freely – or walk away. What this means for art is that it is taken seriously as something which has value, as a product of artifice, but at the same time as something whose usefulness may come to an end (which naturally does not imply the abolition of art, but is an indication of the ultimate point that might be reached in our dealings with art).

The usefulness of art comes to an end, presumably, when the whole cultural spectrum is something in which we can move freely. But the thought that there will ever be such a cultural spectrum would be only utopianism of the crudest sort were it even clear that we should aim for a culture that simply frees (from what? to what?), rather than one that constrains, initiates, teaches, generates distances, provides authorities, is intolerant towards the intolerable, and with its institutions and forms protects from the anarchy of unbridled instinct.

25 I share Peter Fuller's evaluation of Bacon (in his *Images of God*, London, Chatto & Windus, 1985, pp. 63–70). For a considered contrary view, see Jessica Gwynne, 'Pain Displayed', *Salisbury Review*, January 1986, pp. 55–7.

26 In *In Search of Wagner*, London, Verso Editions, 1981, p. 147.

27 *Will to Power* (posthumous), ed. Walter Kaufmann, New York, Vintage Books, 1968, section 827.

28 Cf. Section VII of 'What the Germans Lack' in *Twilight of the Idols*, 1889 and the references to dance in Nietzsche's works, especially *The Gay Science* and *The Case of Wagner*.

29 Anthony Savile in *The Test of Time*, Oxford University Press, 1982, p. 280.

30 Cf. F. R. Leavis, *The Common Pursuit*, Harmondsworth, Penguin Books, 1976, p. 292.

31 R. M. Rilke, Tenth Duino Elegy lines 12–15, *Selected Works*, vol. II, p. 246.

INDEX

www.ingramcontent.com/pod-product-compliance
Ingram Content Group UK Ltd.
Pitfield, Milton Keynes, MK11 3LW, UK
UKHW021956180926
475701UK00003B/53

* 9 7 8 0 4 1 5 7 2 7 8 0 8 *